Fertilizer in America

Fertilizer in America

From Waste Recycling to Resource Exploitation

RICHARD A. WINES

TEMPLE UNIVERSITY PRESS
PHILADELPHIA

To A. Hunter Dupree

Library of Congress Cataloging in Publication Data
Wines, Richard A.
   Fertilizer in America.

   Bibliography: p.
   Includes index.
   1. Fertilizers—United States.   2. Organic wastes
as fertilizer—United States.   1. Title.
S633.2.W75   1985   338.4'766862'0973   84-26855
ISBN 0-87722-374-2

Temple University Press, Philadelphia 19122
© 1985 by Temple University. All rights reserved
Published 1985
Printed in the United States of America

Contents

Acknowledgments

My primary debt of gratitude is to Professors A. Hunter Dupree, John L. Thomas, and Naomi R. Lamoreaux at Brown University. They advised and encouraged me in every way possible and set high standards of scholarship.

The staffs of the Rockefeller Library at Brown, Eleutherian Mills Historical Library, East Hampton Free Library, Maryland Historical Society, Suffolk County Historical Society, Nassau County Museum, Massachusetts Horticultural Society, Boston Public Library, and Baker and Widner Libraries at Harvard have been generous in their assistance and made research at their institutions a genuine pleasure. E. B. Graves of the Agrico Chemical Company and C. P. Harrison of the Tennessee Valley Authority sent me packets of difficult-to-obtain material about the early days of the fertilizer industry. John Wickham, Margaret Rossiter, Vernon W. Ruttan, Harold L. Burstyn, and my mother, Virginia Wines, have offered advice and information. The Fertilizer Institute, New Jersey Historical Society, Woods Hole Historical Collection, Newark Public Library, and Boothbay Region Historical Society diligently answered my queries. Dale E. Baker and Edward H. Quigley made helpful suggestions regarding soil chemistry.

Fertilizer in America

Introduction

As the United States urbanized in the nineteenth century, complex technological and economic systems developed connecting urban centers and the surrounding agricultural regions that supplied them with food, fuel, and raw materials. One of these systems recycled the nutrients from urban wastes back to the countryside as fertilizer to increase agricultural productivity to satisfy the ever-expanding urban markets. This urban-rural nutrient recycling system developed rapidly during the middle decades of the century and resembled the recycling efforts modern ecologists often advocate as the best solution to contemporary environmental and economic problems. By the last quarter of the century, however, the modern chemical fertilizer industry, based on nonrenewable supplies of raw materials and fossil fuels, had largely replaced recycling as the primary source of nutrients for American agriculture.

Although the term "recycling" was not coined until the twentieth century, many nineteenth-century Americans were as enthusiastic about the concept as are contemporary environmentalists. They often used the term "cycle" to describe the reuse of waste material as food for plants. They talked about the "circulation of nutrients," "nature's reciprocity system," the "universal law of compensation," and the "round of creation, dissolution, and reproduction" in the same way that modern ecologists discuss recycling systems. They actively sought ways to convert wastes into fertilizers and other useful products. Many nineteenth-century farmers and agricultural experts were intensely concerned about the serious consequences that might result if the nation did not carefully recycle the wastes of urban consumption.

The development of recycling and the subsequent emergence of commercial fertilizers were closely connected with the rapid growth of East Coast cities during the nineteenth century. New urban markets early in the century provided incentives for many farmers to fertilize their soil.

3

The cities were also, initially, the sources of the wastes that provided most of the fertilizing materials not found on or near the farm. Later, they became the centers for the manufacture and distribution of commercial fertilizers. Throughout the period, they served as centers of the information networks that facilitated the development of fertilizer technology. In return, farmers using commercial fertilizers were able to supply the large quantities of bulky and perishable commodities necessary for urban growth and prosperity.

During the second half of the nineteenth century, increasing urbanization began to undermine the recycling system it had helped create in the first half of the century. First, the continued growth of the urban population created demands for more agricultural produce than could be raised using only recycled wastes as fertilizers. Second, the growing complexity and size of cities made it more difficult to recycle some kinds of organic waste. Finally, the growing commercial and informational networks centered in the cities allowed merchants first to substitute imported Peruvian guano for recycled waste products and later to develop the manufacturing capacity to produce superphosphates and other commercial fertilizers that eventually replaced most fertilizers produced from recycled wastes.

This study began as an examination of an urban-centered technological system. The story of the evolution of the system, however, will take us far beyond the confines of "urban history." Many parts of the story could be considered "agricultural history," since they occur in rural areas around major cities. Moreover, attempts to explore the articulation of the system will lead us even further astray on occasion—to the fisheries along the Maine coast, to phosphate mines in South Carolina, and even to remote guano islands off the coast of Peru and in the mid-Pacific. The full story includes tales of diplomatic intrigue, instances of shameless exploitation of scarce natural resources, and even a nineteenth-century raw-material monopoly similar in many ways to the modern OPEC oil cartel.

This is not a story of straightforward technical improvement. Rather, the development of the fertilizer industry should be considered part of the evolution of a technological system that included not only the production of fertilizers, but also their distribution, their application by farmers, and their utilization by growing plants. The system included agriculturalists, producers, merchants, and other human actors in both urban centers and surrounding agricultural areas; it included production technologies, commercial networks, and agronomic techniques; it also included chemical and biological processes in the production of fertilizers, in the reactions of fertilizers in the soil, and in their absorption by crops.

In this kind of technological system, evolution results from efforts to maintain equilibrium after either internal or external factors have produced instabilities within the system. The system attempts to maintain its stability by making small substitutions. The type of substitution is determined by the system's overall configuration, which is largely preserved. Although the new materials and practices are introduced as equivalent substitutes for materials and practices previously used, the new ones frequently are slightly different and produce unanticipated results. This creates new instabilities that necessitate or allow further small substitutions. Thus, what begins as a minor substitution designed to preserve the existing system may, combined with other similar substitutions, lead to radical evolutionary alterations of the system.

The development of commercial fertilizers is a good example of the evolution of a complex technological system. Although the entire system was immensely complex because of variations in soil, plants, climate, and raw materials, the manufacturing techniques and agronomic practices were simple. It is relatively easy to separate scientific discovery from technological evolution, since there were few important scientific "discoveries" and most of these had been made in Europe and adopted in the United States only after long waiting periods. Fortunately for the historian, the large numbers of participants in the system (including millions of farmers) meant that practitioners and experts tended to communicate with each other through the press. Thus, many of the discussions of the merits of various technological changes, which might be lost in other fields, are preserved here. Nevertheless, this model of the evolutionary development of a complex system, mainly by the process of marginal substitutions, should be applicable to other areas of the history of technology and, possibly, to areas of social, political, and cultural history as well.

CHAPTER I

The Recycling System

The many European observers who traveled through the United States in the late eighteenth and early nineteenth centuries were astonished at the lack of attention American farmers paid to maintaining soil fertility.[1] Since labor was scarce and new land was plentiful, farmers made little effort to use as fertilizers any of the manure and other waste material produced on their farms.[2] In essence, farmers were mining nutrients stored in the soil by the forest ecosystem. In this open-ended system crops removed minerals from the soil but farmers returned nothing to make up the deficit.

At the beginning of the nineteenth century, these wasteful practices were being replaced in a few areas near growing East Coast urban markets by a recycling system in which nutrients removed from the soil by plants were eventually returned to the soil by farmers. American farmers had long been aware of the intensive use of human and animal wastes in parts of Europe and Asia, but the nutrients in the virgin soil had been rich enough and the supply of new land had been plentiful enough to support exhausting farming practices for a century or more.[3] By 1800, however, yields had declined so precipitously in many of the older parts of the East Coast that the stability of the static, precommercial, nonrecycling system was seriously threatened. At the same time, growing urban markets and increased competition from new areas in the West provided additional incentives for the development of a new system of agriculture.

The central element of this new system was the considerable energy devoted to maintaining soil fertility, partly by less exhausting crop rotations, deep plowing, and other improved tillage methods, but principally by recycling as fertilizer all available manure produced on the farm. Maximum manure production necessitated that hay and other crops should not be sold off the farm, because that practice drained the soil of its limited store of fertility. Instead, hay should be fed to livestock on the farm to enhance fertility through the resulting manure. The main feature of this closed cycle system was the recycling of nutrients on the farm as recommended by the old English adage, "keep more livestock to make more manure to raise more hay to feed more livestock, et cetera."

The new system emphasized the conservation of manures and fitted the frugality and self-sufficiency long practiced by most American farmers. Agricultural experts and the new agricultural press repeatedly admonished farmers to save all available manure to increase soil fertility. Farmers were told that manure "should be the grand moving power in the production of an abundant return," "the life of a farm," "the most powerful agent in the hands of a farmer," "the first object of every good farmer" or simply that "manure is money."[4] A correspondent in one of the journals wrote that the farmer's "barnyard is his mine, his manure is his gold dust."[5] Frequent articles in all of the agricultural journals urged the conservation of all possible manure as the key to a farmer's success.[6] These articles often cited Roman, Greek, or biblical precedents and current English or French practices.[7]

Many of the advocates of the conservation of manure recognized it as a cycle. They reasoned that farmers could not continue to take nutrients from their soil and return nothing.[8] They urged farmers to imitate nature which returned decaying herbage and the dung of animals to the soil to nourish future growth.[9] As the concept developed, a correspondent in an agricultural journal described it as the "round of creation, dissolution, and reproduction."[10] Another correspondent urged that anything which had been animal or vegetable matter should be considered as capable of becoming plant matter again.[11] In 1844, Charles T. Jackson, a well-known Boston chemist and geologist, discussed the need of a "continued circulation" of plant nutrients to prevent their loss from the system.[12] Still later, this concept developed into what a speaker to the New York State Agricultural Society in 1859 called a "universal law of compensation" that all organic and inorganic matter used by crops must be returned to the soil, that nothing should be wasted or thrown away.[13]

The concept of recycling may have been more important than the practice of recycling. Farmers and their advisors thought in terms of recycling systems. Even if they did not or could not practice recycling, this was the standard by which farmers judged each other, the assumption being that the larger the manure pile, the better the farmer.[14] Although recycling became a much less important part of the fertilizer system as the century progressed, it remained the ideal of many farmers and agricultural experts and the concept influenced many of the subsequent developments that led to the fertilizer industry.[15]

Local Expansion of the System

Once farmers began to recycle wastes as fertilizer, they sought ways to make the process more efficient. Beginning in the 1830s, agricultural periodicals ran frequent articles on ways to manage farmyard manure.[16] They discussed the best shapes for barns and barnyards to preserve the maximum quantity of manure. They described ways of composting and applying manures and debated whether manures should be "long" or "short" (i.e., fermented or unfermented). Journals often published detailed descriptions of elaborate arrangements made by some of the "best" farmers to conserve their manures. At the same time, observers everywhere chided the majority of farmers for wasting large portions of their available manures.

As farmers appreciated the importance of manure, they attempted to supplement the amount or improve its quality by adding materials available on or near their farms. Following the doctrine that anything that had been life could support life again, farmers began to reach out for other organic materials readily at hand.[17] They carried forest leaves into cow yards and hog pens to absorb the liquid portions of excrements and be converted into valuable fertilizers.[18] Many farmers also used swamp muck and peat for fertilizers, as Samuel Dana recommended in his 1842 *Muck Manual for Farmers*.[19]

Farmers accustomed to using manure also turned their attention toward the use of other kinds of waste produced on a farm.[20] The agricultural press urged farmers to scour their farms for any available vegetable or animal waste. It reminded them that dead animals, liquid drainings from the kitchen and washroom, and most other refuse could be turned into valuable manure if properly preserved. One editor summed up this doctrine by stating "in a word, save everything in the shape of refuse or offal, it is all good to make crops grow."[21] Agricultural improvers such as Samuel Dana provided elaborate recipes for turning all of these materials into valuable composts.[22] One enterprising New Yorker, George Bommer, even attempted to patent and sell rights to a composting method for rapidly converting farm wastes into a "rich manure."[23]

Along the coastal areas of the Northeast, fish and seaweed were logical substitutes for farmyard manure, which seldom could be procured in sufficient quantities. The use of fish was not new, but farmers did not fertilize extensively with fish until the last decade of the eighteenth century.[24] By the first decades of the nineteenth century, farmers on the shores of Long Island, Connecticut, and Rhode Island organized fishing companies and caught immense numbers of menhaden to use directly as a fertilizer or as an ingredient in making compost. Farmers near the shore also harvested considerable quantities of seaweed that they applied directly, composted, or used as a substitute for straw or leaves in the farmyard.[25]

Fish, seaweed, muck, peat, leaves, and farm refuse all fitted into the recycling system because they were organic substitutes for farmyard manure. Along with the efforts to conserve and preserve manures, the use of these materials represented increased applications of human and animal energy to preserving soil fertility. Though not strictly recycling, the use of these organic substitutes can best be considered as a locally expanded version of the recycling system. None of the materials were purchased, they were seldom obtained more than a few miles from the farm, and the farm labor force did most of the fishing, digging, carting, and composting. Thus, all of these improvements reinforced the local recycling system and strengthened the image of that system in the minds of farmers.

New York and Long Island

The booming East Coast cities, such as New York, Philadelphia, Boston, and Baltimore, created strong incentives to expand the local recycling system almost as soon as it was firmly established. The intense market gardening around these cities required far more manure than could be produced on the farm. The demand for hay, the principal energy source for urban transportation systems, created even more dramatic demands for manure. Since hay could not be transported great distances because of its bulk, farms just beyond the market gardening zone began to specialize in hay production for the urban market. This practice, however, directly contradicted the belief that the sale of hay and other bulky crops would inevitably ruin the soil. The logical resolution of this contradiction was simply to enlarge the recycling system: cart the hay and other produce off to cities and return to the farm the manure produced in the consumption of this produce.[26]

The best example of this recycling relationship developed between New York City and nearby agricultural areas of Long Island. Many observers considered the western end of Long Island one of the "gardens of America."[27] Timothy Dwight, after traveling through the area between Brooklyn and Jamaica in 1804, remarked that he could "remember no spot of the same extent where the produce of so many kinds appeared so well."[28] This remarkable productivity was, however, not due to the natural fertility of the soil. As Dwight noted, the soil was sandy and light, although it responded well to manure. Others called the soil "hungry" or "indifferent."[29] Without the application of manure, the fertility of the soil was so depleted in many areas that it could only produce ten bushels of wheat per acre.[30]

The combination of light soil and proximity to the city broke down the local recycling system advocated by the experts. Even if all the hay and produce had been fed to animals, there would not have been enough manure produced to maintain the fertility of a soil which, even when first cleared, was not outstanding. Moreover, the city's profitable market for hay and produce tempted most farmers within reach to keep as few animals as possible and to ship the maximum amount of their produce into the city. These two factors combined to create a demand for manures unequalled anywhere else in the country.

The proximity to New York further stimulated change because the city was the center of a well-developed agricultural information network. Agricultural societies based in the city provided a forum for the discussion and dissemination of information. Both the Society for the Promotion of Agriculture, Arts, and Manufacturers and later the New York State Agricultural Society had many members in the city. The American Institute, which served as the city's agricultural society, was a clearinghouse for information. James J. Mapes was one of its more prominent members. Moreover, leading journals, such as the *American Agriculturist* and Mapes's *Working Farmer*, were published in New York.

In addition, many wealthy city residents, who were exposed to information from Europe and who may have participated in the discussions of the agricultural societies, had country homes on Long Island, where they became "gentleman farmers" willing to experiment and try progressive methods.

Long Island's farmers first attempted to supply the needed nutrients by expanding their fertilizer system locally. They took full advantage of the abundant fertilizing resources available from the sea as substitutes in the manure cycle. During his 1804 trip to the island, Timothy Dwight was amazed at the "almost incredible" number of fish caught and used.[31] In addition, the island's farmers made liberal use of black mussels, seaweed, "horsefeet" (horseshoe crabs), fish offal, salt hay, marsh muck, and leaves as manures.[32] All of this required large amounts of time and energy. Fishing alone consumed forty or fifty days each spring and summer when the farmers prepared and watched the cooperatively owned seines, hauled in the catches, and carted five tons of fish per acre to their fields.[33] Much larger amounts of seaweed or muck were needed. It is not surprising, therefore, that the island's farmers sought easier ways to maintain the fertility of their fields.

For farmers on the western end of the island, the solution was to import the large quantities of manures produced in New York City, which was a veritable manure factory. The large number of horses that powered the city's transportation system produced immense quantities of dung. Some of this accumulated in the stables scattered throughout the city, but much of it was deposited on the streets along with garbage and the excrement of the numerous hogs that roamed New York's streets eating the garbage. Dairies located in the city produced still more, and the human population contributed its share.[34] Various commercial and industrial establishments, such as slaughterhouses, tanneries, soap boilers, and fish markets produced additional waste.

Removal of these wastes was essential for the cleanliness and health of nineteenth-century New York. Fortunately, the agricultural market for manure provided considerable incentive for the owners of dairies and stables to have their manure removed. Sometimes they sold it directly to farmers, but often they sold it to dealers who carted it to the outskirts of the city where they composted it with sawdust, spent tanner's bark, spent charcoal from the rectifying establishments, and other urban wastes to produce a light, friable manure for which farmers were willing to pay premium prices. Although the manure was eventually removed, it still posed numerous problems for the city, since it normally accumulated in large piles near the stables and dairies before it was finally carted away. These piles and the operations of the manure dealers bred flies and emitted noxious odors that elicited frequent complaints from neighboring residents.[35]

Removal of the street dirt presented a more complex problem. Its ingredients—animal dung, garbage, and ashes—made it a valuable manure. Collecting it, however, was a difficult political problem. At various times in the first half of the nineteenth century, the city farmed out the right to clean the streets, hired contractors, or attempted to do the job itself. Sometimes the city received enough revenue from the sale of street dirt to offset most or all of the cost of street cleaning. Whichever method was used, the opportunities for corruption were great and the complaints about dirty, foul streets frequent. Moreover, much of what was carted was placed in huge piles on wharves and empty lots to await eventual sale to farmers, creating further problems for the inhabitants of the city.[36]

Long Island and New York City in essence became a huge recycling system. The island provided energy for the city in the form of food, hay, and fuel wood.[37] The city in turn provided nutrients for agriculture in the manures and wastes which were returned to the island. Often, the very same boats, and later trains, which carried hay, produce, and fuel wood westbound to market, returned loaded with manure or other fertilizers produced or purchased in the city.[38] Reasonably accurate estimates of the volume of this trade are not available for earlier than 1840, but in that year the farmers of Queens County, which then included present-day Nassau County, sold about half of their hay, their principal cash crop, grossing about $280,000. In return, the same farmers purchased about $228,000 worth of manures from the city. In addition, they produced on their farms another 460,000 cartloads of manure, worth as much as the amount purchased from the city. With approximately a half-million dollars' worth of manures, these farmers produced crops valued at almost two million dollars.[39]

This enlargement of the system may have been done consciously. According to the president of the Queens County Agricultural Society in 1848, the farmers of his county had "proceeded for years, upon the principle, that it is better and more profitable to sell the corn and the oats, the straw and the hay, for cash in the market, and with it, to purchase the manures required for the growth of their crops, and the maintenance of their farms in good condition."[40]

Farmers had crossed that important psychological boundary between self-sufficient and capitalistic farming. Instead of attempting to be largely self-sufficient, and selling only what was left over, farmers were gearing their production to urban markets. By purchasing large amounts of fertilizers, farmers had become businessmen, buying raw materials and selling finished products. The editors of the *American Agriculturist* acknowledged this transition later when they described the "long lines of wagons" and the "hundreds of sloops and schooners" that carried manure to Long Island where farmers paid ready cash for it and "manufactured [it] into vegetables, grain, and hay for the City market."[41] This transition from self-contained recycling to purchasing raw materials may have been a more important step toward the commercialization of agriculture than the decision to grow crops for the market, which most farmers seemed to make almost automatically where it was feasible.

Once Long Island's farmers began expanding their system in these ways, they continued to enlarge it further. They brought in stable manure not only from Brooklyn and New York City, but from as far away as Albany.[42] Moreover, once the concept of purchasing manures was established, farmers began searching for other purchased substitutes. Stable manure contained 66 to 85 percent water, but less than one-twentieth the nutrient content per ton of modern commercial fertilizers. Twenty to forty tons were generally needed per acre and the high cost of transporting and applying such bulky material encouraged farmers to seek more concentrated substitutes.[43]

The first substitute for animal manures was leached ashes, which Long Island farmers purchased in large quantities, beginning in the late eighteenth century.[44] These "spent ashes" sometimes came from soap boilers in New York City, Boston, Philadelphia, Providence, and Baltimore.[45] Long Island farmers also sent boats up the Connecticut and Hudson rivers to purchase leached ashes from potash works.[46] The cost of the ashes, delivered to the landings on Long Island, was between 12¢ and 25¢ per bushel in the 1830s and 1840s, although it had been lower earlier in the century.[47] Farmers were willing to pay this high price, cart the ashes up to ten miles and apply between fifty and one hundred fifty bushels (approximately one-and-a-half to four tons) per acre once during the course of each rotation, because the effects on Long Island's soils were so striking. Ashes also fitted into the expanded recycling concept, since they were an organic waste product that could enhance soil fertility. Moreover, the ashes purchased from New York City were produced from cordwood sent in from Long Island as farmers cleared their land. Thus these farmers were recycling back to their soil nutrients from their soil.[48]

Although the results from the first applications of leached ashes were remarkable, the reactions in the island's soil made using ash unstable over the long term. After the first few applications, farmers noticed that subsequent applications had little effect on their crops.[49] Probably leached ashes functioned mainly because the lime they contained decreased the acidity of the soil, thus increasing the availability of whatever nutrients were present. Once those nutrients were exhausted, however, subsequent applications of ashes, without replenishment of the basic plant nutrients, had no further effect.[50] Leached ashes may also have been effective because of the small amounts of potassium that remained after most of the potash (K_2O) had been leached out at the ashery.[51] If potassium were deficient, as was likely on the light sandy soils of the island, applications of that element would have temporarily increased productivity until some other essential plant nutrient, such as phosphorus, was exhausted.[52] Then, subsequent applications would have had little effect.

Although the use of ashes was not a permanent solution to the island's fertility problem, trade in ashes opened commercial distribution networks and accustomed farmers to using such materials on their land. Moreover, once farmers began using supplemental fertilizers, it was difficult for them to return to their old ways without drastically cutting productivity. Thus, when the efficiency of ashes declined, farmers sought substitutes among the waste products of New York City.

One substitute suggested by the recycling mentality was ground bones. The best-informed agriculturalists were aware of their successful use in Great Britain since the early part of the century. Bones were available in large quantities as refuse at urban slaughterhouses. They fitted the recycling concept since using bones as a fertilizer returned to the soil nutrients that had originated in the soil. Bones also fitted into the commercial network already supplying fertilizing materials to the island's farmers. Bones, moreover, were an attractive substitute for ashes, because only ten to twenty bushels were needed, about one-fifth the amount of ashes typically used. Consequently, the island's farmers began to use considerable quantities of bone in the 1830s and 1840s.[53]

Bones, however, were not a chemically equivalent substitute for leached ashes. Bones were valuable mainly because of their high phosphorus content; ashes had been valuable because of their potassium and lime. As happened frequently during the evolution of the fertilizer system, the choice of possible substitutions was governed by the mentality of recycling, not by the discoveries of analytical chemists. The success or failure of these substitutions was determined by the information farmers gathered from observations of their crops, not from experiments of agricultural researchers.

The extensive recycling system between New York City and Long Island paved the way for additional substitutions in the 1840s and 1850s. The limited quantities of stable and street manure available, the difficulty of transporting such bulky material, and the cost of ashes and bones, encouraged farmers to seek alternative fertilizers.[54] Not surprisingly, some of the first fertilizer manufacturing concerns were established in or near New York City where they could take advantage of the distribution networks and the demand for commercial fertilizers established by ashes, bones, and manures purchased from the city. These companies, which produced a fertilizer called "poudrette" from night soil, assumed that their principal market would be Long Island.[55] When guano was introduced in the mid-1840s, Long Island was one of the first areas to adopt that concentrated but expensive commercial fertilizer.[56] Similarly, several of the first superphosphate companies in the 1850s were on western Long Island (in Brooklyn) and across New York Harbor in New Jersey.[57]

By midcentury, Long Island's fertilizer system was the inverse of the original recycling ideal. Both economic opportunity and the stern necessity imposed by poor soil had contributed to this change. The information network composed of agricultural societies and journals centered in New York City facilitated it. The fertilizing materials available from New York allowed it. Instead of keeping as much of their hay and produce on the farm as possible, to feed the stock and produce manure, these farmers were selling large quantities of bulky, exhausting crops in urban markets and bringing back large quantities of city manures to maintain soil fertility. As these farmers drastically altered their practices, they began to reverse their attitude of self-sufficiency. They no longer considered the best farmers to be those who produced the most manure, but those who purchased the most manure.[58] Long Island's farmers were so imbued with this new mentality and so committed to the enlarged recycling system, that when the federal census began collecting fertilizer statistics in 1880, they were some of the largest purchasers of commercial fertilizers in the country.[59]

What happened on Long Island also happened in other areas close to growing urban centers. Westchester, Richmond, and Dutchess counties in New York and nearby areas of New Jersey all experienced some of the same effects from New York's growth as did Long Island.[60] The same phenomenon occurred in the areas around Philadelphia, Baltimore, Boston, and other smaller East Coast cities.[61] In each case, the surrounding areas shared the benefits of urban markets, manure supplies, and information networks, but the strength of the recycling network varied according to the soil type and the size of the city. The larger the city and the weaker the soil, the stronger the recycling relationship.

Philadelphia and Its Hinterlands

The relationship between Philadelphia and its surrounding agricultural areas in southeastern Pennsylvania and southwestern New Jersey was similar in many ways to that between New York City and western Long Island. As on Long Island, long cultivation had so depleted the soil that wheat yields of less than ten bushels per acre were common.[62] To alleviate this problem, an urban-rural recycling system, similar to the one around New York, developed between Philadelphia and its adjacent areas. Farmers carted hay and produce into the city and returned with stable manure and street sweepings for their fields.[63] Like New York City, Philadelphia not only offered its hinterland attractive markets and abundant supplies of fertilizing wastes, but also served as a center of information. The "gentleman farmers" of the region, who in 1785 formed the Philadelphia Society for Promoting Agriculture, were connected through the city with the latest developments in England and on the Continent.[64]

One of the most striking results of this relationship was the introduction of gypsum (also called land plaster or plaster of Paris) into the area. Richard Peters, the first president of the Philadelphia Society for the Promotion of Agriculture, began using gypsum a few years before the Revolution, after having seen it used by a Philadelphian who supposedly learned about it in a letter from friends who had observed its use during a trip to their native Germany.[65] Peters and his friends in Philadelphia, including Benjamin Franklin, actively promoted the use of gypsum after the Revolution.[66] By the late 1780s, at least a few farmers in each of the counties surrounding Philadelphia were using gypsum.[67] The first cargoes came from France, generally as ballast, but before the end of the century Nova Scotia replaced France as the main supplier.[68]

Gypsum appears to have been a major exception in the recyling system. It was an imported mineral, not a recycled organic waste, although some farmers did attempt to substitute it for recycled wastes from the farmyard or city. Gypsum, however, reinforced the recycling system on the naturally fertile limestone soils of southeastern Pennsylvania because it allowed the growth of clover and grass as part of an improved rotation. This in turn allowed the feeding of more livestock, thus producing more manure to maintain the fertility of the soil. In those areas of Chester and Lancaster counties with the best soils, the use of gypsum and later limestone enabled the establishment of a highly efficient on-the-farm recycling system.[69] The fertility-enhancing ability of this system was augmented because cattle could be cheaply purchased from western parts of the state and fattened in the area for the Philadelphia market.

Gypsum may, however, have had a greater effect on the thinking of agriculturalists than on the agricultural productivity of Pennsylvania. Only small amounts were used, often only two or three bushels per acre, compared with twenty or more cartloads of manure.[70] Once it became part of the accepted practice of Pennsylvania farmers, however, it helped prepare the way for the introduction of other purchased fertilizers, although the pathway was not a direct one.

The infatuation with gypsum lasted barely thirty years in most areas of southeastern Pennsylvania.[71] Its remarkable initial effects were probably due mainly to a shortage of available calcium or sulfur in the soil. Repeated applications, however, did not have the same effect.[72] Apparently, with adequate supplies of calcium and sulfur, the exhaustion of some other essential nutrient became the limiting factor in plant growth.[73]

At this point, most farmers substituted lime for gypsum. A few agriculturists in the area had used lime since the seventeenth century and its use had a long history in Europe, dating back to Roman times, but large numbers of farmers did not begin to use lime as a soil amendment until after gypsum began to lose its power.[74] In making the substitution, farmers undoubtedly thought they were substituting a similar substance and accomplishing the same result, since both gypsum and lime contained calcium, had similar appearances, could be applied in the same way, and neither was an organic waste. However, the two substances operated in different ways chemically. While the principal function of gypsum was to supply calcium and sulfates and possibly to improve the soil texture, the principal function of lime was to decrease the acidity of the soil, thus temporarily increasing the availability of many plant nutrients. Since the nutrient supply was not permanently increased, the use of lime alone would have quickly exhausted the soil and productivity would have again plummeted. On the naturally productive soils of Lancaster and Chester counties, however, the lime, like gypsum, allowed the production of sufficient clover and grass to feed sufficient livestock to produce sufficient manure to maintain and even enhance the natural fertility of the soil and reinforce the local recycling system.[75]

Since limestone was available locally in many parts of southeastern Pennsylvania, its use did not create the commercial network or the commercial mentality that characterized the urban-rural recycling system around New York. Lime could be quarried, carried a short distance to the farm, and burned in kilns with wood cut locally. Total carting distance seldom exceeded ten miles.[76] At the kilns, burnt lime cost about eight to ten cents per bushel (about $3.60–$4.00 per acre) in the 1820s.[77] Later, when lime burners substituted nearby deposits of anthracite coal for wood, the cost of burnt lime was even less.[78] Since lime was a locally obtained material, it did not create a commercial network as had the use of ashes and bones on Long Island. Moreover, since it successfully reinforced the local recycling system, it did not create the need for substitutes. Therefore, farmers in these areas of Pennsylvania were slow to take up the use of guano in the 1840s.[79] As an 1853 report to the Pennsylvania State Agricultural Society indicated, Chester County and Delaware County farmers used lime widely with marked effects, but used guano and superphosphates only "to some extent."[80] By 1880 farmers in southeastern Pennsylvania still purchased only moderate amounts of commercial fertilizers compared to farmers near other East Coast cities.[81]

The fertilizer system developed quite differently in southwestern New Jersey near Philadelphia than it did in southeastern Pennsylvania. The main difference was the soil type. The New Jersey soils were light, sandy loams rapidly exhausted by cultivation, but that responded well to the application of fertilizers. Since the situation was very similar to that of Long Island, it is not surprising that the evolution of the two systems followed parallel courses. The New Jersey farmers began by recycling street sweepings and stable manure when their own manure supplies proved inadequate. They then substituted ashes purchased in Philadelphia and lime purchased at kilns further up the Delaware. Eventually they substituted guano and other commercial manures into the system until, by the 1880s, they became intensive users of commercial fertilizers.[82]

Some New Jersey farmers slightly farther removed from Philadelphia attempted to reinforce the recycling system locally. A band of greensand marl (glauconite), a soft, recent sedimentary deposit with significant amounts of potassium, stretched across the state from Monmouth County southwest to Delaware Bay. Its utility was known in a few areas by the beginning of the century, but it was not widely dug until the 1820s. Then pits were opened where farmers dug the marl and carted it up to ten or fifteen miles to their farms. The results on the barren, sandy soils of the region were dramatic.[83]

Although marling with greensand functioned as part of a local closed recycling system much as liming did in Pennsylvania, several factors conspired to encourage New Jersey's farmers to expand their system by substituting commercial fertilizers for marl when the former became available. Since the best marl contained only about 1 percent phosphoric acid and 6 percent potash, between one and twenty tons were needed per acre.[84] Since this was not a complete fertilizer because it contained so little phosphoric acid and no nitrogen or calcium, it was necessary to supplement the marl with other fertilizers and lime. Most importantly, because the thin, sandy soil of New Jersey required large amounts of fertilizers, farmers who tried to use marl to reinforce on-the-farm recycling, as their Pennsylvania neighbors did with lime or gypsum, found the greensand marl could not sustain sufficient fertility to produce the required fodder to feed enough livestock to fuel a manure cycle. The difficulty of the New

Jersey farmers was increased because they were not as well located as their Pennsylvania brethren to purchase cheap cattle from the West to fatten for the Philadelphia market. Therefore, instead of reinforcing the local closed recycling system as lime did in Pennsylvania, greensand marl served as a pathway for expanding the fertilizer system in southern and central New Jersey, as had happened on Long Island. Although the use of marl initially slowed the introduction of guano, by 1880, central and southern New Jersey farmers used relatively large amounts of commercial fertilizers while farmers in Lancaster and Chester counties in Pennsylvania used only moderate amounts.[85]

Baltimore and the Chesapeake Region

As on Long Island, in the Chesapeake region a light, "hungry" soil and the proximity (augmented by excellent transportation systems) of a major metropolitan market encouraged farmers to seek ways to increase productivity.[86] Although the improvement did not begin in the area as early as on Long Island and although Baltimore, the region's urban center, could provide neither markets nor manure supplies on the same scale as New York, the Chesapeake area developed a fertilizer system that closely resembled that of Long Island, and Baltimore eventually became the center of the emerging fertilizer industry.

The soil of the Chesapeake region was in even worse condition in the early nineteenth century than that of Long Island. Travelers and local observers described large areas as "barren," "dreary," or as having a "miserable aspect."[87] Long cultivation of exhausting crops, especially tobacco, had worn down the soil to the point that in many areas it was simply abandoned and allowed to grow up in scrub.[88]

Many agriculturalists in the area understood what was happening. One observer in the first volume of the *American Farmer*, the Baltimore-based journal in the forefront of the movement for agricultural improvement, decried the cultivation of tobacco, which did not provide "either food for the soil or for men."[89] Farmers were aware that, instead of recycling nutrients, their system simply bled the soil. However, with so much land, such low yields, and the small numbers of animals typically kept, the possibility of improving land by growing more hay, feeding more animals, and producing more manure, was very slim.[90]

The on-the-farm recycling system, as envisioned and to some extent practiced in the North, was not practical here. Richard Parkinson, an English traveler who toured the area at the turn of the century, calculated that under the best of conditions a four hundred acre farm could produce only enough manure to fertilize the garden, a few turnips, and perhaps another three acres per year.[91] At this rate, 133 years would have been required to manure the whole farm. Since each field needed to be manured at least once every five or six years to maintain fertility, the task was hopeless.

Farmers in the area were aware of the advantages of recycling but lacked the manure supplies to make the system work. Therefore, farmers who desired to improve their land had to turn to other materials locally available. Those near the numerous waterways used fish and seaweed.[92] They also carted muck from the many marshes.[93] The principal resource they used as a substitute for the insufficient manure was marl, a soft sedimentary material, in this area valuable principally for its high calcium carbonate (lime) content and available on or near many farms.[94]

Marl had long been known in a few localities, such as Yorktown, but its extensive use had to wait for the impetus provided by expanding markets for wheat in the 1820s and 1830s.[95] Edmund Ruffin, with his famous works on calcareous manures, was only the most prominent of the many advocates of marling.[96] Throughout the area, in the 1820s and 1830s, farmers began to seek out marl deposits and cart it to their fields.[97]

The sea, marshes, and marl pits, combined with whatever could be gleaned from the compost pit and farmyard, were the basis for the first steps in the improvement of Maryland and Virginia agriculture. All of this material was obtained locally, and was seldom carted more than a few miles. Indeed, given the large quantities needed per acre, longer cartage would have been economically prohibitive. Since these materials were obtained locally, they functioned, like fish on Long Island, as local supplements to the recycling system, even though the recycling system itself was more an ideal than a reality in the region.

The leading advocates of marl considered it an adjunct to the recycling system. Ruffin makes clear in his *Essay on Calcareous Manures* that marl primarily functions by making more available to plants the "putrescent manures" (vegetable and animal wastes) supplied to the soil by the decay of vegetation or applied artificially as yard manure, muck, or mud.[98] Thus, like lime and gypsum, marl facilitated the recycling system. Farmers who had attempted to increase the fertility of their fields by using composts and other recycled manures were often the first ones to try marl.[99] One apparently typical farmer on Maryland's Eastern Shore wrote a letter to the *American Farmer* in 1829 describing extensive use of marl and composts as if the two were intimately connected in his mind. Even the use of the term "calcareous manures" shows that Ruffin and others considered marl a substitute for organic manures.

The use of marl, muck, seaweed, and fish presented the area's farmers with a number of difficulties. First, these materials were unavailable in many areas. Second, because of their low analysis and high bulk, they were expensive to transport. Third, marl, the most widely used of these, was not a complete fertilizer. The long-term use of marl, like lime, initially increased but ultimately decreased the fertility of the soil. Thus farmers using it had several incentives to seek better substitutes.

In the areas immediately adjacent to cities, urban-rural recycling systems developed as they had around New York and Philadelphia to make up the deficit in locally available fertilizing materials.[100] Baltimore's gentleman farmers, periodicals, and, after 1846 its agricultural society, like those of New York and Philadelphia, disseminated information about methods of agricultural improvement.[101] In addition, the same type of recycling system that had developed in the North, involving rural production, urban consumption, and the return of waste to the soil, developed around the Chesapeake cities.[102] The practices of Horace Capron, one of Maryland's leading improvers in the 1840s, embodied the essence of this expanded recycling system. As he cleared his farm near Laurel, Maryland, he sold cordwood in nearby Washington, D.C., and in return, purchased large quantities of the city's ashes to apply to his depleted soil.[103]

The fertilizing resources and the growing markets of the Chesapeake cities such as Baltimore and Annapolis led to the development of hay and truck farming in the 1830s and 1840s.[104] Improvements in transportation further enlarged the urban-rural recycling system as truck farming spread to the vicinity of Norfolk, Virginia, where in 1856 Frederick Law Olmstead found numerous intensively cultivated farms scattered among the old worn-out ones. Not only did Norfolk farmers sell their produce in Baltimore and other cities further north, but they also brought back manure from those cities.[105] Wherever farmers engaged in this type of intensive agriculture, the recycling mentality was strongest.

The Chesapeake cities later became the centers for improvements.[106] When guano and superphosphates were introduced, it was farmers connected to the urban recycling system who first experimented and then enthusiastically adopted those materials. Since they were already in the habit of purchasing fertilizers and using them on a large scale, when a material became available that appeared to be more powerful and at the same time easier to use, they quickly took it up.[107]

In those parts of the Chesapeake region further removed from the urban centers, farmers who found it costly to use marl or muck, because of the large amounts needed or lack of local supplies, sought commercial substitutes.[108] Ashes were available only near cities and towns. Elsewhere, in the 1820s and 1830s, the only available substitute was lime that allowed the growth of clover to recycle as green manure.[109] To obtain lime and gypsum, farmers had to enter the commercial network not only to sell products, but to purchase raw materials. Once they were accustomed to purchasing one type of soil amendment—like farmers near New York and Philadelphia—those in Montgomery County, Maryland, and similar areas were well prepared to substitute bone and guano into their fertilizer system when these newer materials became available in the 1840s and 1850s.[110]

The Southeast

South of Virginia, the development of the fertilizer system was very different. Since the Southeast had few major cities, southern farmers did not benefit from the stimulus of urban markets, the information from urban communication networks, and the abundant supplies of recyclable wastes generated by urban communities. The extensive farming practices and the lack of barnyard-kept livestock on many southern farms, moreover, made the development of the local recycling system unlikely. As long as uncleared land existed in the Southeast and more fertile new land was available in the Southwest, planters had little incentive to improve their lands, except in specialized areas along the Atlantic coast.[111] In the 1840s, most parts of the Southeast still used little manure.[112] Thus, neither the recycling system nor the recycling mentality developed in the area. Without these, the introduction of guano, superphosphates, and other commercial fertilizers later in the century was much slower than in the North, although the region's soil badly needed fertilizer and the region eventually became the leading consumer of commercial fertilizers.[113] Even when commercial fertilizers were belatedly introduced into the South, they were not part of the recycling system as in the North. Instead, as some southern editors complained, farmers in that region often used commercial fertilizers as an alternative to recycling the valuable fertilizers produced on their own farms.

CHAPTER II

The Impact of the Recycling Mentality

The recycling mentality, firmly in place by the 1840s, suggested a wide variety of different directions in which the system of fertilizer production and use could evolve. Some of these pathways led to immediate dead ends, others developed for a few decades and then vanished, and still others, by process of substitution, led to the modern system of widespread use of commercial fertilizers manufactured from inorganic raw materials.

The recycling concept led many farmers and agricultural experts to explore every possible recyclable waste. The *American Agriculturist* asked in 1847, "How many substances do we see lying about the country which might be gathered up and applied as manure, thus removing many an intolerable nuisance, and at the same time greatly increasing our crops?"[1] In 1853 the *Plough, Loom, and Anvil* stated that "manure worth more than one hundred millions of dollars is annually thrown away in the United States."[2] Among the substances considered by agriculturalists were: refuse salt from saltworks; horn shavings, piths, and bone dust from comb and button manufacturers; gas lime and ammoniacal liquor of the gashouses; blood and offal from slaughterhouses; hair, flesh scrapings, and spent tanbark from tanneries; spent bone charcoal from sugar refineries; distillery refuse; the waste of glue factories; cottonseed; sawdust; anthracite coal ashes; wool rags; flax waste; lime from paper mills; brine and salt from country stores; and human excrement.[3] Not surprisingly, most of these waste materials were produced by urban processing or consumption. Most of these materials were logical substitutes for already popular fertilizers, but many did not succeed because they were not chemically equivalent, such as the anthracite ashes some farmers attempted to substitute for wood ashes. Other attempted substitutions failed because materials such as pickle brine or horn piths were not available in large enough quantities to become significant.

A few of the substitutions suggested by the recycling mentality led to important new fertilizers. Blood and slaughterhouse offal became primary sources of nitrogen for the fertilizer industry, once techniques were developed to dry and process the material. The ammoniacal liquor from gasworks, converted into ammonium sulfate, provided another source of nitrogen. Manufacturers of the first complete fertilizers in the 1850s mixed the ammonium sulfate, dried blood, and offal with ground bone, phosphatic guano, or superphosphates. These recycled urban waste materials continued to provide the nitrogen for mixed fertilizer until artificial sources of nitrogen were developed in the late nineteenth and early twentieth centuries.

Another line of development began with the spent bone charcoal from sugar houses. Bone charcoal or bone black was used to clarify the sugar syrup in the refining process. It was frequently reburned and reused several times, until it was no longer capable of drawing impurities out of the sugar. It did, however, retain all of the phosphorus that made bones a valuable fertilizer. Because this spent bone charcoal was available at the refineries in Boston, New York, Philadelphia, Baltimore, and elsewhere in large quantities at very low prices, several of the early superphosphate manufacturers used this as their first raw material. Later, when supplies proved inadequate, manufacturers easily substituted other sources of phosphate in their superphosphates.

The Case of Bones

The introduction of bones provides an excellent example of how the recycling concept led to the development of a new fertilizer. The early advocates of the use of bones clearly thought in terms of recycling wastes. In 1837 the editor of the *Cultivator* explicitly compared bones with other vegetable and animal wastes farmers were returning to the soil as fertilizers.[4] The following year, the *Cultivator* reprinted from a British journal an article that described bone as part of the "great circle of revolving nature" in which the phosphorus and other nutrients contained in bone could be returned to the earth to produce new vegetation.[5] Another editor stressed the importance of returning the nutrients in bones back to the fields from which they had come.[6] One member of the American Institute in New York City illustrated this nutrient cycle, which he considered part of a "divine economy," with the observation that "the bones left on the field of Waterloo were gathered up to be put on the corn and grass fields of England to make other bones for the fields of Sebastopol and Balaklava [sic]."[7]

Some of the authors writing in the agricultural press about the use of bones thought not only in terms of urban-rural recycling systems, but in terms of national ones. They argued that the exportation of bones to Great Britain in the 1830s and 1840s was robbing American fields of nutrients to enrich the soil of Great Britain. They urged that the only way to prevent irretrievable decline in national productivity was to return the nutrients in bone to American fields.[8] Horace Greeley, who was alarmed by the gradual impoverishment of American soil as exhausting cereals were exported to Great Britain, placed these arguments in a nationalistic context. He managed to incorporate the need for recycling into an argument for protecting American industry with tariffs to create better markets in the United States for agricultural products so that more of the wastes could be recycled onto American fields.[9]

The success of the first small-scale experiments with bones created a demand for bone from farmers near cities, who were already purchasing fertilizers such as ashes, lime, or city manures. Raw whole bones, however, are not very effective. Bones must be ground to make the phosphates, the most valuable ingredient, readily available to plants. An even better method is to dissolve bones in sulfuric acid to produce "superphosphate," a process which had been pioneered in Great Britain in the early 1840s.[10] Initially, farmers following the recommendations of the agricultural journals attempted to grind or acidulate the bones themselves, but the necessary grinding machinery was expensive and the sulfuric acid was difficult to obtain and dangerous to use.[11] Thus, bone dealers not only had to collect bones from butchers and bone boiling establishments, but also grind and process the bones. By 1833, two mills were already operating on Long Island, where the urban-rural recycling system was most intensively developed.[12] By 1841, bone mills also had been set up in the vicinities of Albany, Boston, Troy, and Baltimore.[13] In the 1850s, several large establishments in the New York area manufactured bone dust and advertised their product widely.[14] One of these, Lister Brothers of Tarrytown, later moved to New Jersey and became a leading producer of agricultural chemicals in the last quarter of the century.[15] Several other bone dealers also built lead chambers to manufacture sulfuric acid and then began to produce superphosphate commercially.

The substitution of materials like bone, which required processing, produced an expanded and more complex recycling system. Once entrepreneurs had invested in manufacturing plants and developed markets for their products, however, the limited quantities of bone available forced them to actively seek other sources of phosphate. Phosphatic guanos discovered on some Caribbean and central Pacific islands were easily substituted in the process. So were rock phosphates, which were discovered in South Carolina in 1867.[16] Thus the recycling concept that first suggested the agricultural use of bones led ultimately to the modern superphosphate industry based on mineral phosphates. Although this new system grew out of recycling, it has ceased to be circular. Nutrients are not recycled back to the soil, but instead allowed to flow through the system and out the end as nonreclaimed wastes.

Night Soil

Of all the urban waste materials considered as potential fertilizers, night soil attracted the most attention, although eventually no viable method was found for its use. Articles abounded on the merits of human excrement as a fertilizer.[17] Chemical analyses showed it to be a valuable source of most nutrients needed for plant growth.[18] The recycling mentality and the analogy often made between food and fertilizers led many observers to speculate that since man ate the richest food of all the animals, his excrement must be better than that of other animals as food for plants.[19] To Lemuel Shattuck, sanitary commissioner of Massachusetts in 1850, it seemed a "law of nature" that such material should be returned to the soil as fertilizer.[20]

Since many of the agricultural writers lived in or near the larger cities, they were all too familiar with the terrible health hazards and esthetic problems posed by then current methods of human waste disposal. By the 1840s scientists were accumulating evidence that human wastes were fouling water wells and endangering public health.[21] The introduction of adequate public water supplies and the advent of the water closet compounded an already difficult problem.[22] The ideal solution seemed to be to solve two problems at once—to reduce the stench and health hazards of the cities and at the same time provide a much-needed source of fertilizers for farmers.

The potential gains from recycling human waste appeared enormous. In 1842 a correspondent of the *American Agriculturist* calculated that the human manure wasted annually from the 350,000 residents of greater New York could produce four million bushels of wheat.[23] In 1844 the *Agriculturist* calculated that just the human urine produced in New York City was worth $350,000 annually.[24] In 1867, one concerned Massachusetts resident calculated that the fertilizing materials wasted in Boston were enough to fertilize 30,000 acres of poor land while the waste of New York City was worth $5,475,000 per year.[25] In 1855 George E. Waring's popular textbook estimated the value of night soil lost annually in the entire country was fifty million dollars, nearly equal to the entire federal budget.[26] In 1871 an advocate of sewage use calculated that 80 percent of the value of human food was thrown away as waste.[27] American and British authors frequently noted that use of this material could make their nations self-sufficient in fertilizers.[28]

Agricultural and sanitary experts proposed a variety of elaborate schemes to recycle this wasted wealth. Editors borrowed some of the ideas directly from the British press, where an even more vigorous discussion of the situation was proceeding in response to that nation's pressing urban problems and need for fertilizers.[29] American editors also often cited practices in Europe and China.[30] In the 1830s and 1840s most of the proposals involved ways to collect, deodorize, and solidify the content of urban privies and cesspools to make an easily transportable concentrated fertilizer. One plan recommended the use of double, removable barrels under every privy.[31] Another plan proposed the establishment of companies to supply households with tight boxes filled with charcoal or other absorbent material that would be placed under the privy openings. The companies would exchange new boxes for filled ones every month and sell the contents of the filled ones to farmers as manure.[32] One advocate of this earth closet plan calculated in 1847 that a company could afford to pay New York City $100,000 per year and still make a profit.[33] Another correspondent to the *American Agriculturist* suggested saving the urine separately and mixing it with sawdust to make a portable manure.[34] In 1871, Horace Greeley still recommended the widespread adoption of earth closets as the best way to prevent the steady diminution of the fertility of American soils.[35] One journal even suggested constructing a pipeline which could carry urine directly from the "convenience" behind New York's City Hall to "North River" docks, where tightly closed boats would carry it to rural regions to use as a fertilizer or as a constituent in compost.[36]

The only scheme along these lines that reached the commercial stage was the manufacture of "poudrette" by deodorizing night soil, a method pioneered in France early in the nineteenth century.[37] By the 1830s, one manufactory at Montfaucon, near Paris, reportedly converted two hundred cartloads of cesspool contents a day into valuable manure while other nearby establishments made a product called "urate" from human urine.

The first poudrette company in the United States was formed in 1837 in New York City by D. K. Minor, who was connected with the *New York Farmer*, the journal of the New York Horticultural Society. The Albany *Cultivator* called his plan a "commendable undertaking," noting that such establishments had already been set up near Paris and London.[38] The manure produced by the company was to be divided among the stockholders, many of whom were Long Island farmers. The company survived for a few years, but never achieved much success.[39] In 1839, however, a partner broke away and set up his own company on the banks of the Hackensack River in New Jersey.[40] It was first called the New York Urate and Poudrette Company, but when the company was incorporated the following year, the name was changed to the Lodi Manufacturing Company, after the town in which it was located. This company survived and continued to produce poudrette and related products for over thirty years.

Unlike its parent company in New York, the Lodi Company was a commercial operation. It offered poudrette on the market for 40¢ per bushel.[41] However, the company encouraged farmers to purchase shares by a unique dividend arrangement. Each farmer who purchased one of 500 special reserved shares for $100 each would receive for five years an annual dividend of fifty bushels of poudrette, which the promoters calculated was the equivalent of a 20 percent yield on capital.[42] The stock may have sold poorly, since five years later the company was still advertising part of the same 500 shares.[43] By 1845, however, the company claimed capital of $75,000.[44]

During the next two decades, the Lodi Manufacturing Company held exclusive contracts with New York City to remove all the night soil collected from the cesspools, vaults, and privies of the city by "scavengers" who were compelled by city ordinance to dump the contents of their carts into the company's boats docked at night in the city.[45] It is unclear how large a portion of this kind of waste the company recycled. Probably the majority of it was not collected and most of what was collected by the scavengers was allowed to spill on the streets (making their carts lighter and their work easier) or was dumped into the water at the numerous wharves around Manhattan, as had been the previous practice, creating a foul mess that had to be periodically dredged by the city to keep the slips clear for commerce.[46]

Each morning the company's boats carried the night soil to the Hackensack River plant where, by a secret "chemical process," the night soil was deodorized and turned into a dry powder.[47] Probably the principal ingredient in the process was the peat, obtainable in large quantities from the 400 acres of meadowland the company owned adjacent to the factory.[48] The factory itself was little more than a large mixing vat, surrounded by fourteen 100' x 12' drying floors with movable roofs, a few storage buildings, and a machine house with "horsepower" for grinding the peat. (See 1863 engraving of plant in figure 2-1.) It was a transitional operation, halfway between farm and factory. The original plan was not only to manufacture poudrette and urate, but also to operate a farm that would complete the fertilizer-to-energy cycle by converting the poudrette into hay for the New York market.[49]

Generally poudrette was substituted into the existing urban-rural recycling system around New York. Advertisements frequently compared poudrette with stable and street manures, claiming that it was so much more powerful that only 170 pounds were needed per acre, thus drastically reducing transportation costs.[50] Actually, several tons probably were needed per acre, but this was still less than the twenty tons of stable manure typically applied.[51] Most farmers who tried poudrette were

Fig. 2-1. Plant of the Lodi Manufacturing Company on the bank of the Hackensack River in New Jersey. In the foreground is one of a series of reservoirs for holding the night soil after it was raised from barges docked adjacent to the small crane. In the center are the drying beds where the night soil was dried in thick layers and mixed with muck dug from the marshes beyond. On the far side of the drying beds are the screening buildings where the poudrette was thoroughly mixed and screened. The other buildings were used for storage, packing, and manufacturing wooden barrels. The buildings on the far right may have been residences for some of the 100–300 workman typically employed. (Engraving from *American Agriculturist* 22 [1863]: 169.)

already purchasing manures from the city. Generally they thought of poudrette as a less expensive substitute for stable or street manures they had been using.

The total poudrette production was neither large nor economically significant. In 1853, the Lodi plant produced only 10,157 barrels, enough to fertilize barely 5,000 acres at minimal rates of application.[52] By the end of the decade, production had expanded to only 60,000 barrels per year.[53] Farmers may have been turned away by advertisements making exorbitant claims that it could ripen crops two weeks earlier, could completely prevent cutworms, and was ten to fifteen times more powerful than stable manure.[54] Many farmers reported successful experiments with poudrette, but a significant number reported failures.[55]

Although the manufacturers claimed poudrette was ten to fifteen times as strong as good stable manure, farmers complained that they found it otherwise in practice because it was diluted with peat or other earthy substances.[56] When Samuel W. Johnson, Yale's agricultural chemist, analyzed the Lodi poudrette, he found it almost worthless. He concluded that despite the "extravagant and persistent claims" for the product, he could not recommend it since it was mostly sticks, coal dust, and other equally worthless material.[57] Although poudrette fit the recycling system in theory, the manufactured product did not meet expectations, either because of difficulties inherent in the process or because of willful adulteration.

A number of companies in other urban areas attempted the manufacture of poudrette, but most of these amounted to little. In 1840, the *American Farmer* suggested a poudrette manufactory should be set up in Baltimore on the New York model.[58] Nothing apparently came of the suggestion until ten years later when Thomas Bayles advertised a "poudrette and bone dust establishment" there.[59] In the 1840s there were reports of poudrette establishments in Philadelphia and Richmond, and as many as ten plants may have operated in the New York area.[60] One of these was established by George Bommer of New York City, who had previously tried to exploit his patented method of making compost.[61] By the early 1850s, however, the only plants still operating in the country were those of the Lodi Company in New Jersey, Bayles in Baltimore, the Liebig Manufacturing Company in East Hartford, Connecticut, and a company in Philadelphia.[62]

Even though many farmers were disillusioned with poudrette, enough of them purchased the product to provide a continuous market through the 1870s. In 1850, the Lodi Company advertised a "new and improved poudrette."[63] In 1857, the company again advertised a "new improved poudrette." They claimed to have been producing the improved poudrette for two years, but it probably was a reaction to S. W. Johnson's critical analysis the same year.[64] In addition, the company offered in 1856 a new product called "tafeu," supposedly the Chinese word for prepared night soil. The initial advertisements claimed it was 95 percent pure night soil "manufactured without any adulteration whatever."[65] Subsequent advertisements claimed, however, that it was three-fourths night soil and one-fourth Peruvian guano.[66] (Regular poudrette was supposedly two-thirds night soil and one-third "vegetable fiber.") In the next decade, the company made further improvements. In 1866, they advertised a "bone tafeu" made from bone, dried night soil, and guano.[67] In 1869, they advertised a "double refined" poudrette which they continued to offer through the 1870s.[68]

In the late 1850s and 1860s, several other companies entered the poudrette and tafeu business. The Brooklyn Fertilizer and Manufacturing Company advertised it had a ten-year contract with the City of Brooklyn to remove night soil, blood, and butcher's offal. From this material the company manufactured an "ammoniated tafeu."[69] In 1863, the Ricardo & Company's Excelsior Poudrette and Fish Guano Works on Staten Island offered a fertilizer made from the night soil and butcher's blood of Brooklyn, Williamsburg, and Green Point.[70] In 1869, the Bromophyte Fertilizer Company of St. Louis, Missouri, attempted to market a similar fertilizer.[71] In the late 1870s, the Baltimore Health Department offered its "Maryland Poudrette."[72] Poudrette production probably peaked in the 1870s or 1880s, and then dropped rapidly to insignificant levels in 1900.[73]

Although poudrette production was relatively small and most of the manufacturers did not last long, many agriculturalists were familiar with the product, either from personal experience or from the large notice it received in the press. For some it was the first processed commercial fertilizer they purchased. Although poudrette was not an entirely satisfactory fertilizer, it did lead many farmers later to purchase other concentrated commercial or "artificial" fertilizers such as guano or superphosphates. When guano was introduced, it was sometimes presented as simply another form of poudrette that had been recycled by a more roundabout method.[74] Promoters reasoned that the same wastes which could be made into poudrette generally washed into the oceans instead, where they nourished the marine organisms that became the food of the fish consumed by the birds that produced guano.

While farmers who used poudrette were likely later to substitute concentrated fertilizers such as guano and superphosphates, the manufacturers of poudrette were not able to make similar substitutions in their part of the system. The manufacture of poudrette from night soil was a waste processing system similar to the manufacture of superphosphates from bones. Unlike superphosphate producers, however, who were able to substitute other raw materials when bone became too scarce or expensive, poudrette manufacturers could never escape from the problems and limitations inherent in their original raw material. Since they could not find similar raw materials to substitute in the manufacturing process, this branch of the fertilizer industry ceased to evolve and died altogether when the supply of night soil disappeared in the last decades of the century.

Sewage Recycling

By the late 1840s, increased water supplies had created sewage disposal problems that were more pressing than the disposal of the contents of urban privies. Since the problem was far more urgent in Great Britain, that nation's chemists, farmers, political economists, engineers, and physicians led the way in attempts to use sewage for fertilizer. Edwin Chadwick, Britain's leading sanitary reformer, discussed sewage farming in 1842.[75] In subsequent years, interested experts prepared elaborate schemes for using the sewage of London and other large cities to irrigate and fertilize large areas of surrounding countryside.[76] Costly experiments were carried out and joint stock companies were formed to take advantage of the opportunities, but little came of all this.[77]

American agriculturists took a keen interest in the various British proposals and experiments, and American agricultural journals were not far behind in discussing this subject.[78] In the early 1850s, James J. Mapes's *Working Farmer* proposed several schemes. One involved purifying New York's sewage by upward filtration in large cisterns filled with charcoal. The sewage would be introduced alternately into the bottom of each of the cisterns and clear water would emerge from the top. The solid matter and nutrients trapped in the charcoal could then be removed and sold as fertilizer.[79] Another scheme proposed using sewage from Newark and Elizabeth to fertilize and irrigate 10,000 acres of drained New Jersey marsh.[80] Mapes also supported plans of John Randall and other sanitary engineers to use New York's sewage for agriculture.[81]

Perhaps the most feasible plan was George E. Waring's 1867 proposal to build a receiving sewer which would skirt Manhattan and then tunnel under the East River to pick up Brooklyn's waste. From there it would proceed to the barren sandy lands on the south side of Long Island, which could be made productive. Waring's plan was based explicitly on a scheme to extend London's sewerage system forty-one miles eastward to Essex, where the sewage could be used to irrigate and fertilize large areas of barren sands and reclaimed land.[82] Six years later, Waring further explored the feasibility of sewage farming while on a trip to Great Britain. His reports on British efforts were a bit less sanguine than his earlier proposal, but he remained committed to the concept.[83]

Waring's career demonstrates how the recycling mentality associated with agriculture influenced thinking about sewage wastes. Before becoming a sanitary engineer, Waring had been mainly interested in agriculture. He had been a pupil of "Professor" James J. Mapes, a leading proponent of waste recycling and the editor of *Working Farmer*; served as an assistant editor of that paper; lectured before farmers' clubs on improved methods; managed farms owned by Horace Greeley and Frederick Law Olmstead; and wrote a popular series of articles on practical farming for the *American Agriculturist*. He had even written a textbook for "young farmers" expounding on the merits of the recycling system and calling night soil "the best manure within reach of the farmer."[84] Farm drainage was one of the subjects he had written about as an agricultural expert. Gradually he extended his field of expertise to include urban drainage and sanitary engineering. Perhaps the transition was sparked by his appointment in 1857 as drainage engineer for Central Park. In the 1870s and 1880s, as a "sanitary engineer" he designed the sewerage systems for a number of North American cities and wrote several books on sewerage and sewage disposal. Then in 1895 William L. Strong, the reform mayor of New York, appointed Waring commissioner of street cleaning for the city.[85] Throughout his career, Waring remained committed to the reuse of urban wastes as fertilizer. His numerous writings and those of his mentor, James J. Mapes, influenced many of his contemporaries, including Horace Greeley.

Conclusion

Interest in recycling human excrement as fertilizer continued throughout the century. In 1871, Horace Greeley wrote that it was "plainly absurd to spend ten thousand dollars for [guano] when [New York] or any other great city annually poisons its own atmosphere and adjacent waters with excretions which science and capital might combine to utilize at less than half the cost."[86] Two years later a *Scientific American* article stated that "it is no exaggeration to say that the problem of conversion of the excremental waste of towns and people and the refuse of factories into useful materials is now engaging as much attention of intelligent minds throughout the world as any social question."[87] The article described six different methods of converting sewage into forms suitable for agricultural use. In 1877, Boston appointed a commission with two engineers to investigate sewage use.[88] In 1880, entrepreneurs attempted to form a "Sanitary Fertilizer Company of the United States" to undertake the conversion of night soil into fertilizer by the earth closet method.[89] Later in the century, the Chemical Division of the United States Department of Agriculture made systematic studies of the use of sewage.[90] Little came of the proposals because of the economic and engineering difficulties involved and because medical experts became more aware of the health risks.[91] But the continuing interest in using urban wastes demonstrates the persistence of the recycling mentality.

The proponents of these various plans never forgot that they were attempting to set up recycling systems. As John A. Dix, future governor of New York, noted in an 1859 speech to the New York State Agricultural Society, towns drew largely on the fertility of the countryside for their subsistence, but gave back little in return. He looked forward to the time when the great cities, instead of draining into the oceans and rivers the remains of what they consumed, would gather up and restore them to the earth.[92] On both sides of the Atlantic, farmers and interested experts continued to regard the sewage of towns as the "natural pabulum" for agriculture, but they were forced to conclude that until satisfactory ways were found to recycle these materials, farmers would have to resort to artificial fertilizers.[93]

CHAPTER III

Guano

The introduction of Peruvian guano into the American market in the 1840s inaugurated the transition from bulky recycled fertilizers to modern commercial fertilizers manufactured from nonrenewable natural resources. Like many technological changes, guano was adopted rapidly because it fitted easily into the existing technological system. Along the East Coast, many farmers perceived guano as a similar substitute for the recycled urban wastes they were already purchasing and using, such as stable manure, ashes, and bones, hence they were willing to try it. Initially, guano was adopted successfully only in those areas where the urban-rural recycling system had accustomed farmers to purchasing commercial fertilizers but could not provide adequate supplies. Once farmers tried guano, they discovered, as the ancient Peruvian proverb stated, that "guano, though no saint, works many miracles." It proved to be a marvelous fertilizer and demand increased rapidly.

Although guano was initially substituted into an existing technological system, it quickly transformed the entire system. It established new sources of nutrient supply, it led to the entry of new classes of entrepreneurs, it changed the way farmers used fertilizers, and, most important, it changed the way agriculturists thought about fertilizers. Although the main period of extensive Peruvian guano use was confined to the 1850s, it nevertheless exerted a significant impact on agricultural practices and the emerging fertilizer industry for the rest of the century.

Early History

The introduction of guano into the American market in the 1840s was not the result of recent discovery or invention. Rather, it followed a long period of information accumulation. Guano, which is the Spanish version of the Inca word "huano," had been known by agriculturalists in present-day Peru at least as early as the second century B.C.[1] After the Spanish conquest, information about guano's marvelous fertilizing powers traveled back to Europe through books such as José de Acosta's *Historia Natural y Moral de las Indias*, published in 1604, Garcilaso de la Vega's *Comentarios Reales que tratan del origen de los Yncas*, published in 1609, and Amédée François Frézier's *Voyage to the South Sea*, published in 1717.[2] The Spanish explorer Antonio de Ulloa visited the coast of Peru during his expedition between 1735 and 1746 and soon published a vivid description of the "prodigious quantities" of guano carried off from the coastal islands to fertilize maize and other crops on the mainland.[3] Despite the wide circulation of some of these accounts, the references to guano by seventeenth-century chroniclers and eighteenth-century travelers had no impact on the European agricultural or scientific community, although, after the commercial introduction of guano in the 1840s, promoters often cited these early authorities to give their product a proper lineage.[4]

Baron Alexander von Humboldt, the German scientist and explorer, brought fresh reports of guano to Europe in the early nineteenth century. Humboldt and the French botanist Aime Bonpland had traveled widely in Central and South America between 1799 and 1804.[5] Among the many specimens the two explorers took back to Europe was a small sample of guano, which they transmitted for analysis to the French chemists Antoine Francois de Fourcroy and Nicolas Louis Vanguelin, whose analysis, published in the *Annales de Chimie* in 1806, confirmed that guano was a valuable fertilizer.[6] This analysis, and another by the German chemist Martin Heinrich Klaproth, brought guano to the attention of the European scientific community. Abraham Rees's forty-one volume *Cyclopaedia of Arts, Sciences and Literature*, the first American edition of which was published between 1805 and 1825, contained an article on guano citing these analyses and quoting heavily from a letter Humboldt had sent Klaproth describing the use of guano in Peru and speculating on the origins of the substance.[7] John Claudius Loudon's *Encyclopaedia of Agriculture*, published in 1825, and Andrew Ure's *Dictionary of Chemistry*, published in 1821, had articles on guano based on the same sources.[8] Sir Humphrey Davy's pioneering work, *Elements of Agricultural Chemistry*, published in 1813, also cited Fourcroy, Vanguelin, and Humboldt in addition to reporting an experiment he had conducted on a sample sent from South America to the British Board of Agriculture in 1805.[9]

In the 1840s, most writers on the subject credited Humboldt, who had become a popular figure in the United States, with having introduced guano to Europe.[10] Humboldt, however, had not considered guano important enough to mention in the widely read account of his travels.[11] Moreover, he apparently had not visited any of the guano islands, although he may have passed several during his return voyage from Callao to Guayaquil in March 1803.[12] The publication of Fourcroy and

Vanguelin's analysis and references in several convenient works kept the subject alive through the 1820s. This information, however, did not lead to any further experimentation other than a few small experiments carried out at the behest of Sir Joseph Banks on Saint Helena in 1811 using guano found on the island.[13]

The first sample brought into the United States was delivered to John S. Skinner, the editor of the Baltimore *American Farmer*, by his son-in-law, a Midshipman Bland, in 1824.[14] Skinner used the two casks that Bland brought ("with some other curious articles" from a trip to South America) for small experiments, and gave a sample to Dr. Julius Ducatel, a "Professor of Agricultural Chemistry," for analysis.[15] Skinner published an account of this "celebrated manure used in South America" in the 24 December 1824 *American Farmer*. This article included references to its use in Peru, and quoted from Fourcroy and Vanguelin's analysis of Humboldt's sample, Rees's *Cyclopaedia*, Ulloa's *Voyage to South America*, and Davy's *Agricultural Chemistry*. Apparently Skinner did not even publish the results of his experiments, and guano dropped from public notice, although it was mentioned in the agricultural press in 1825 and 1831 and small quantities may have been imported in 1830 and 1832.[16]

Nearly twenty years elapsed between Skinner's first article about guano and its successful introduction as a commercial fertilizer in Great Britain in the early 1840s and the United States a few years later. Guano caught on first in Great Britain because that nation's farmers already had developed a system of agriculture that not only intensively utilized manures produced on the farm and purchased from urban centers, but also, by the 1840s, required the importation of large quantities of bones to use as fertilizers.[17] Guano fitted easily into this system. Farmers who were already accustomed to purchasing organic fertilizers were willing to try guano when the Liverpool firm of W. J. Myers and Company imported twenty casks of it in 1840 for experiments.[18] The dramatic results from the first experiments created a brisk demand for guano in Great Britain. The return of political stability in Peru facilitated the guano trade, which quickly reached large proportions.[19] In 1841, seven vessels carried 1,733 tons to Great Britain, the following year forty-four vessels carried 13,094 tons there, and the amounts increased rapidly in subsequent years.[20]

The success of guano in Great Britain was directly responsible for the introduction of guano as a commercial product in the United States. Numerous British newspapers, scientific journals, and advertising pamphlets soon described successful experiments with guano.[21] The American agricultural press, which always closely followed developments in Great Britain, relayed news of these experiments to American farmers. For instance, in 1842 the Albany *Cultivator* carried two articles on guano which it called the "most powerful natural manure known."[22] Samuel Dana's *Muck Manual for Farmers*, published the same year, included a description of guano and copied an analysis made by a European chemist.[23]

Descriptions of British experiments must have aroused some interest among a few American farmers and scientists. In 1844, the *American Agriculturist* predicted that if guano could be exempted from duty and imported at a reasonable price it would find considerable markets on the Atlantic shore.[24] In the next few years several small amounts were imported for experiments, the most extensive of which were carried out between 1843 and 1845 by the Boston horticulturist James E. Teschemacher using guano samples brought by Capt. John Percival of the United States Navy.[25]

The first commercial shipments of guano arrived in December 1844 when the British firm of Anthony Gibbs and Sons sent to their agents, Samuel K. George in Baltimore and Edwin Bartlett in New York, two shiploads of guano totaling 700 tons.[26] The recent discovery of a deposit of guano on the African island of Ichaboe had caused a temporary glut on the British market.[27] Gibbs and Sons, the exclusive consignee of the Peruvian government for the exportation of guano, apparently sent the two shiploads to the United States to see if an alternative market for guano could be created there. To promote this guano, the American agents published a pamphlet in January 1845 summarizing experiments in Great Britain and the United States, recounting the history of guano, and giving instructions for its use.[28] Interest generated by this publicity and by previous articles was great enough that half of the first 700 tons was sold within four months, but demand grew slowly, and two years passed before the initial shipments were completely sold.[29]

Substitution of Guano for Recycled Fertilizers

Guano caught on rapidly in the United States in the mid-1840s primarily because the fertilizer system had developed in such a way that guano could be easily substituted for materials which were in use, but were not entirely satisfactory or were too expensive. During the previous two decades, farmers in the vicinities of Baltimore, New York, Philadelphia, and other East Coast cities had gradually developed new systems of agriculture which involved intensive application of fertilizers. These farmers had first used locally available materials such as yard manure, muck, marl, or fish, but increasingly those within convenient transportation distance of cities had begun to purchase city manures, lime, ashes, and bones. Thus these farmers were accustomed to purchasing fertilizers, and urban-centered distribution systems that could accommodate guano when it was introduced were already in place.

The introduction of guano was facilitated by its similarity to many of the fertilizers farmers were using. Like the stable and street manures, guano was an excrement. Because of its high ammonia content, guano even smelled like the animal manures farmers were accustomed to using.[30] Writers described guano as a "manure" and compared it with other well-known manures. For instance, Bartlett's 1845 pamphlet claimed that one hundred pounds of guano was the equivalent of five or six tons of barnyard manure, or about one hundred times as strong.[31] Authors less interested in selling guano still considered it thirty to fifty times as effective as stable manure, four to fourteen times as concentrated as horse manure, and ten times as powerful as fish.[32] All of these writers stressed (correctly) that guano was similar in composition and capabilities to well-established fertilizers, but much more concentrated.

Guano first became an important fertilizer in areas such as western Long Island, where an extensive urban-rural recycling system had already developed but where even the tremendous fertilizing resources of New York City were insufficient to maintain the productivity of the island's light soil. In 1845 the *American Agriculturist* suggested that because it was "so cheap" guano might be the best way to rebuild the depleted fertility of Long Island's soils.[33] Experiments had begun on the island in 1843 when a New York lawyer applied guano to the garden at his Flushing country estate.[34] The following year, the *Cultivator* reported several other experiments on western Long Island.[35] By 1847 a correspondent reported to the *Cultivator* that "large purchases" were being made for the season in Queens County.[36] By 1852, according to the *Cultivator*, Long Island was ahead of all northern sections in the use of "this great fertilizer."[37] Similar patterns of guano use developed around other urban centers where waste recycling systems had evolved. In 1848 A. B. Allen, a New York dealer in guano and other agricultural supplies, reported that "for the past five years it has been very extensively used and highly approved of by farmers and gardeners in the neighborhood of Boston, New York, Philadelphia, Baltimore, Norfolk, Charleston and New Orleans."[38]

Guano established itself most rapidly in the Chesapeake region where worn-out soil responded marvelously to its application and the commercial fertilizer systems which had grown up around the area's cities and towns facilitated its introduction.[39] Some of the first experiments were conducted in 1844 near Petersburg, Virginia, where one farmer figured guano was cheaper than either bones or purchased stable manure.[40] The Petersburg *Intelligencer* recommended that "the present price of guano places it within the reach of all who are in the habit of purchasing town manures."[41] Other experiments were carried out the same year in Montgomery County, Maryland, where extensive use of purchased fertilizers had become an established practice.[42]

Generally, farmers substituted guano for recycled urban wastes they had been purchasing. Teschemacher correctly assumed in 1845 that guano would compete with city manures.[43] Farmers substituted guano for ashes and bones.[44] In addition, they explicitly compared it with poudrette, the first (supposedly) concentrated manure that many northern farmers had purchased.[45] Long Island, the principal market for poudrette, also became the principal northern market for guano. Inevitably, the farmers who were farming most intensively and who were purchasing the largest amounts of city manure, bone, ashes, and lime, were the first to purchase guano. For example, in Maryland, Horace Capron and Edward Stabler, leading agricultural reformers who had been large purchasers of ashes, bone, and lime, were among the first to try guano.[46]

Guano would not have been substituted into the system if the existing fertilizing materials had been adequate. However, stable manures were too bulky to be carted far and poudrette had failed to live up to initial expectations.[47] Ashes and lime were effective for only a few applications unless supplemented with other fertilizers and bone was scarce and expensive. Guano was much cheaper to transport than bulky fertilizers, was readily available, and provided much larger amounts of nitrogen than any of the fertilizers it replaced. Thus, even though guano cost much more per ton than any other common fertilizer, since smaller amounts were needed per acre, it was less expensive for most farmers to use. Thus guano helped preserve the stability of agricultural systems constantly threatened by soil exhaustion.

The introduction of guano was made easier because its use was compatible with the recycling mentality that had been developing since the beginning of the century. Many agricultural editors and farmers considered the use of guano as simply the expansion of the recycling system. They thought of guano as an indirect way to return the nutrients in urban wastes to the soil. According to this line of reasoning, if the nutrients in urban wastes were not returned to farmers' fields as city manure, bones, ashes, or poudrette, but were instead washed into the oceans, the nutrients then supported the growth of plankton and other marine plants that became the food for the fish that were eaten by the seabirds that produced guano. Importing guano and applying it to the soil completed the cycle. Writers explicitly described this as a larger-scale cycle.[48]

For agriculturalists who thought in terms of nutrient cycles, the importation of guano made up for the loss of fertilizing materials lost by erosion, waste, or the exportation of crops. J. S. Skinner, who as editor of the *American Farmer* had been the first to suggest experiments with guano in 1824, wrote later that "it would seem that Providence designed the ocean to restore to the earth, some portion of the riches carried by rains and rivers from the sides of the one into the depths of the other."[49] The *Working Farmer* considered guano the "natural method" to replace nutrients carried off in crops or lost through erosion and the "surest and most direct course to national wealth."[50] Daniel Lee, the editor of the

Southern Cultivator (Augusta, Georgia), showed a similar appreciation of the nutrient cycle when he noted in an 1853 article that since every stream running into the sea carried "some of the earthy constituents of crops," man must go there for "his richest and cheapest manure."[51] Some agriculturalists thought that the importation of guano was the inverse of the exportation of commercial crops. They argued that the exportation of crops such as cotton, wheat, or tobacco must be balanced by the importation of fertilizers such as guano, otherwise the nation's natural bank of fertility would be depleted.[52]

Because guano was so compatible with the existing fertilizing system and the mentality of farmers, it was not the radical innovation it would have been twenty years earlier. Farmers still considered guano an innovation, and initial experimentation was cautious. But because of the directions in which the fertilizer system had evolved, what could not have been accepted in 1824 was readily accepted in the mid-1840s.

The Patterns of Adoption

In 1845 the *American Agriculturist* feared there would be "a perfect mania" for guano.[53] Within three years this prediction became a reality. From 700 tons in 1845, the importation of guano grew to 1,013 tons in 1848, 21,243 the following year, and peaked at 175,849 tons in 1854, enough to fertilize approximately one million acres at the low rates often applied.[54] The attention paid to guano by the agricultural press mirrored and to some extent increased the growing interest in guano among agriculturists. In the early 1840s, agricultural journals contained only a few scattered references to British experiments. By the middle of the decade, however, those same journals carried frequent articles on guano. The widely read *American Farmer* had no references to guano in 1840, one in 1841, twenty-seven in 1845, and eighty-five in 1851. Other journals followed the same pattern, although the numbers of articles were less.[55] In the beginning of the decade, most of the articles were accounts of British experiences, often clipped directly from British journals. By the end of the decade, however, most of the accounts were of American origin, again reflecting the growing rate of experimentation and use in the United States.

Where the fertilizer system was conducive to the introduction of guano, individuals of ample means in or near the large cities made the first experiments, then other agriculturalists less able to assume the risks of the initial experiments observed the results and tried guano on a small scale themselves.[56] Gradually, the use of guano spread to the majority of farmers in the area. Montgomery County, Maryland, provides a good example of how the use of guano spread. The leading agriculturalists of the county, including Edward and Caleb Stabler of Sandy Springs, conducted the first small-scale experiments in 1844 and 1845.[57] Guano was a logical substitute for the large amounts of bone and lime they were already purchasing. According to Horace Capron, a leading agricultural reformer from neighboring Prince Georges County, by 1847 the first Montgomery County experiments were "infecting the whole neighborhood."[58] That same year the *Georgetown Advertiser* reported that the people of Montgomery County seemed "to have become entirely convinced of the great advantages of using guano on their soil." The paper added that guano had introduced a "new era" into agriculture by its "magical influence" on the soil and had "instilled into the farmers a spirit of improvement."[59] That year, the county's farmers purchased at least seventy tons of guano.[60] Two years later, one Montgomery County farmer reported that the effects of applying guano on worn-out land were so favorable that they created "almost as great a fever to procure it" as then afflicted many seekers of gold in the West.[61] By 1853 farmers in the area considered guano so "indispensably necessary" for wheat and corn that they purchased fifteen hundred tons, enough to fertilize 150,000 acres.[62] The same year, Edward Stabler claimed that guano had increased the wheat crop 200 percent since 1845 and had a "magical" effect on land previously considered worthless.[63]

After the early experiments in Montgomery County and a few other parts of the Chesapeake region, the use of guano spread rapidly until, by 1852, it was in common use in Maryland on both sides of Chesapeake Bay and was spreading rapidly in Virginia.[64] The widespread use of guano markedly increased the productivity of Maryland and Virginia farms and led to large increases in agricultural land values.[65] The agricultural press reported that yields increased so greatly that some sharp individuals were able to purchase "worn out" land, apply guano, and pay for the land, guano, and other costs from the proceeds of the first crop.[66] In some areas, guano virtually transformed the landscape, turning large areas of waste where only sedge and pine bushes had grown into productive farmland.[67]

The adoption of guano was especially rapid because the results of experiments often appeared to be better than they were. Since guano is an unbalanced fertilizer disproportionately high in nitrogen, the initial visual effects of an application was striking.[68] Leaves immediately became dark green and foliage grew rapidly. On the basis of this initial impression, farmers concluded that the effect produced by guano was astonishing.[69] Because of the low percentage of phosphorus, actual yields of grain often were not as large as initial observations led farmers to believe they would be. Since yields were generally only estimated or roughly measured, these shortfalls were often overlooked.[70] Without a doubt, guano was an effective fertilizer, but the dramatic initial impression caused by its high nitrogen content hastened its acceptance. The effects appeared so dramatic that proponents attributed the few experimental failures to extraneous factors.[71]

The adoption of guano was hastened by the information farmers received from the agricultural press and pamphlets distributed by guano merchants. Both the journals and the pamphlets contained accounts of experiments by farmers. Since initial impressions, not scientific measurements, were the basis for many of the reports, the exaggerations of firsthand observations were perpetuated. Some promoters and users considerably overestimated the powers of guano. A few even claimed that guano could control bugs, grubs, and rot.[72]

Farmers also had access to scientific information in the agricultural press and promotional materials. Some experiments and analyses had been done independently by scientists, but many of the analyses were done for and financed by guano merchants who used the results to promote their product.[73] Scientific findings and recommendations seldom initiated interest in guano use. They did, however, often confirm the firsthand observations of farmers, and provided scientific explanations for those observations. The typical promotional pamphlet had sections on the "testimony of science" and the "testimony of practice," but generally farmers paid far more attention to the longer sections on the "testimony of practice."[74]

Although the availability of information influenced the rate at which the use of guano spread, the system of fertility maintenance already in place in an area ultimately determined the likelihood that guano would be successfully introduced into the area. Wherever recycled or purchased fertilizers were not already used, guano was not adopted. Guano spread most rapidly in the worn-out soils of the Chesapeake area, where improved systems had been established but available fertilizers were inadequate. It also spread rapidly on Long Island, where an intensive commercial fertilizer system had been established, although not quite as rapidly as in the Chesapeake, since the regional recycling system between New York City and western Long Island could supply a larger portion of the fertilizers needed than could similar systems in the Chesapeake.[75] The use of guano spread far less rapidly in areas where the existing practices were adequate to maintain stability. Thus Pennsylvania, where local lime deposits allowed production of enough manure to maintain fertility, and central New Jersey, where greensand marl did the same, were relatively slow to take up guano.[76]

Ironically, Georgia and the Carolinas, where the gap between fertilizer need and supply was wider than anywhere else in the country, were slow to experiment with guano, even though they eventually became the largest users of concentrated fertilizers.[77] The soil of the principal cotton-growing areas in the Piedmont region of Georgia and the Carolinas had never been as fertile as the rich black soils further west and was further depleted by several decades of cotton cultivation.[78] Recycling systems, however, had not developed in the South as they had in the North, partly because of the lack of urbanization. Moreover, many planters moved to the more fertile land readily available in the Old Southwest, rather than invest the time and labor necessary to maintain or improve the fertility of their farms in the Southeast. Not until the 1850s, when new land became scarcer, did farmers remaining in the region begin to seek ways to improve their soils.[79]

Then, after the dramatic success of guano had been demonstrated elsewhere, the consumption of guano began to grow rapidly in the Southwest.[80] A few planters in the region had experimented with guano in the 1840s and early 1850s, mainly on the coastal islands where the poor, sandy soils and the long cultivation of the highly profitable sea island cotton had forced them to establish local recycling systems using mud, marl, and farmyard manure.[81] In the late 1840s, the *Southern Agriculturist*, published in Charleston, South Carolina, showed much less interest in guano than its counterparts in the mid-Atlantic and northern states.[82] Farmers in the interior of Georgia did not begin to experiment with guano until 1852 or 1853.[83] In 1859, use was still so light that a Sparta, Georgia, farmer still had to order his guano (and all other fertilizers) directly from Baltimore.[84]

Finally, in the last years of the decade, farmers in the Southeast began to purchase larger quantities of guano. The Augusta, Georgia, *Chronicle* remarked in 1860 that the use of guano was then growing rapidly in Georgia and the Carolinas.[85] The Central Railroad of Georgia reported that it carried almost four times as much fertilizer in the first half of 1860 as it had in all of 1859.[86] Although the Civil War interrupted the process, the Southeast eventually became the leading consumer of commercial fertilizers in the postbellum years.

The Supply of Guano

Although guano was initially merely a substitute for apparently similar fertilizers already in use, it quickly led to new commercial networks for the supply of fertilizers, modified agricultural practices, and changed the way farmers thought about fertilizers. This new technological system encompassed guano digging on islands off the coast of Peru, the sailing ships that carried the guano to East Coast ports, the distribution channels to the farmers, and agronomic practices.

The Chincha Islands were the principal source of guano during the first decades of worldwide guano distribution.[87] (See figures 3-1, 3-2, and 3-3.) These three islands, the largest less than one mile long and one-third mile across, lie about ten miles offshore from the Peruvian port of Pisco. Before the international guano trade began in the 1840s, the islands had

Fig. 3-1. View of the Middle Island and roadstead at Chincha Islands, 1868.
(From *American Agriculturist* 27 [1868]: 20.)

Fig. 3-2. Guano bed of North Island in the Chincha Islands, 1868. (From
American Agriculturist 27 [1868]: 20.)

been uninhabited, except by large numbers of cormorants, pelicans, gannets (also called piqueros or boobies), and other guano-producing birds.[88] These birds flourished on the islands because the upwelling from the bottom associated with the cool, north-flowing Peru Current (also called the Humboldt Current) brought nutrients close to the surface of the Pacific Ocean near the islands, allowing a rich growth of plankton and other microorganisms. Countless anchovies fed on the plankton and in turn became the food of the incredibly large flocks of birds that roosted on the predator-free islands and produced the guano.[89] Since the Andes Mountains protect the Peruvian coast from rain, the valuable nitrogen content of the bird excrement was effectively preserved in the dry guano deposits two hundred feet deep in many parts of the islands.[90]

After Great Britain and the United States began importing large quantities of guano, the islands became centers of ruthless exploitation of both human and natural resources. At the peak of production in the late 1850s, several thousand soldiers, government officials, and laborers swarmed over the islands. Within two decades, they had managed to

Fig. 3-3. Ship loading guano in the Chincha Islands about 1853. (From *National Magazine* 3 [1853]: 553.)

exhaust most of the available supplies of guano. Working conditions were terrible, both because of the inhospitable nature of the islands and the utterly relentless way in which the laborers were driven. Initially, slaves were used until Peru abolished slavery in 1855. Then the operators imported Chinese coolies, but continued to treat them like slaves. Conditions were so bad that many of the laborers choose suicide as their only escape.

Ship captains found the situation offshore equally difficult. Loading operations were inefficient and sometimes dangerous. Ships were often delayed several months as several hundred merchantmen waited at anchor for their turn at the loading chutes. Peruvian officials were high-handed, expected numerous bribes, caused many unnecessary delays, and generally made life difficult for the captains. Guano was an obnoxious cargo, difficult to load, and its high density tempted captains to take on more than safety rules allowed. Altogether, this was a thoroughly unpleasant trade.[91]

The guano trade rapidly assumed large proportions. In 1841, only a few thousand tons were shipped to Great Britain, but by 1852, 220,000 tons were exported to the world market annually.[92] Exports grew during the ensuing decade. By the 1870s, over ten million tons of guano worth half a billion dollars had been removed from the Chincha Islands.[93] Great Britain was always the largest customer, accounting for over half of total exports, the United States received about a quarter of Peru's guano exports, and the rest went mainly to continental Europe and the Caribbean.

The impact on Peru of massive guano exportation was so large that a recent historian of that nation has called the period from 1840 to 1875 Peru's "Guano Age."[94] Guano was the country's most important export. Not only did the government's profits, which were over half the proceeds from guano sales, stimulate the economy through salaries, graft, and public works, but additional income came to Peru from purchases made by the guano fleet and bribes to Peruvian officials.[95] Guano was also a major factor in Peru's internal politics and international relations, as both domestic political factions and foreign nations attempted to control the wealth produced by the guano islands.

From the beginning, the guano trade was tightly controlled by the Peruvian government, which relied on guano as its principal source of income. The government never felt it had the resources to run the trade directly. Instead it usually consigned the guano exports to exclusive agents who handled the digging, shipping, and sale of the guano. The government paid the consignees for their expenses and paid them a commission based on those expenses and a percentage of the profits.[96] Under these arrangements, the Peruvian government assumed the entrepreneurial risks, but it also guaranteed for itself most of the profits, which were generally large since the government exercised monopoly control over the world guano trade.[97] When the contracts were signed, the consignees always made large loans to the Peruvian government, to be repaid from the proceeds of the guano sales. In effect, the consignees acted as "bankers for the Republic."[98] This arrangement gave Peru access to immediate credit enjoyed by few other nations.

In 1847, after a series of short-lived agreements with various consignees, the Peruvian government made the London firm of Anthony

Gibbs and Sons exclusive agents for most of the world.[99] Gibbs and Sons continued as agents for Great Britain through the 1850s, but the United States trade was consigned in 1852 to the Peruvian firm of F. Barreda and Brother for five years.[100] The firm was well-connected politically in Peru. One senior member of the firm was the brother-in-law of Don Juan Ignacio de Osma, the Peruvian chargé in Washington.[101] Other high Peruvian officials were also allegedly connected with the monopoly.[102]

The consignees usually appointed exclusive agents to handle guano sales. The first American agents of Gibbs and Sons in 1844 were Samuel K. George, a Baltimore dry goods merchant, and Edwin Bartlett, a New York merchant.[103] After 1852, the new consignees, F. Barreda and Brother, functioned as their own wholesale agents in Baltimore.[104] The Barreda brothers moved to Baltimore and conducted the main part of their trade there, but in 1856 removed to New York.[105] This second move was partly in response to Maryland's recently passed guano inspection law, but New York may have been a more logical center for imports since it could generate more exports to Peru.[106] The Barreda brothers remained exclusive agents throughout the peak of guano imports in the mid-1850s until 1858 when a new consignee returned the bulk of the trade to Baltimore.[107]

The consignees handled all shipping and insurance arrangements. They engaged ships on a charter basis, generally paying from twelve to fifteen dollars a ton in the early 1850s.[108] This was enough to make the charters attractive to shipowners despite the long delays and high costs associated with loading the guano in Peru, since guano provided a convenient return cargo for American ships which had carried wheat and manufactured goods from East Coast ports to the Pacific coasts of North and South America.[109] Initially, many of the finest ships afloat were involved, but as the trade aged in the 1860s and 1870s, the ships sent for guano were aging hulks whose owners may have been more interested in collecting insurance money than charter fees.[110]

A network of agents and dealers developed to distribute the guano. The consignees handled the wholesale business themselves or through exclusive agents in the principal ports of entry. The consignees and their exclusive agents occasionally sold directly to large farmers, but generally they sold to dealers. These dealers sold to smaller country merchants and to farmers at a markup of two dollars or more per ton. The agents and dealers were usually established dry goods merchants in large cities, such as Samuel K. George in Baltimore, or agricultural editors and publicists, such as Samuel Sands, the publisher of the *American Farmer,* and Solon Robinson, who was connected with the *American Agriculturist.*[111] A. B. Allen of New York was both a publisher of the *American Agriculturist* and the proprietor of a recently founded agricultural warehouse that supplied seeds, tools, and other materials needed by farmers.[112] As the guano trade grew, some of the merchants, such as William Whitelock and Robert Turner of Baltimore, began to specialize in guano.[113] Ideally this urban-centered network of dealers and merchants distributed the guano to the farmers and provided needed credit. When guano was in short supply, however, as it was in the late 1840s and early 1850s, dealers bought up all available supplies and then charged what the market would bear, sometimes as much as 36 percent above the already high official price.[114] Farmers complained vociferously about this practice until F. Barreda and Brother took over the agency in Baltimore in 1852 and promised to keep the market better supplied so as to prevent hoarding by merchants.[115]

At the local level, the merchants typically were not specialized. Country stores added guano to their stock of merchandise and handled it in the same way they did salt or cloth. Other local merchants simply added guano to their existing business as did an Oyster Bay, Long Island, lumber yard operator who sold not only guano but also lumber, coal, and poudrette.[116]

Baltimore was the center of the guano trade from its inception, except for the years from 1856 to 1858 when the consignees imported most of their guano through New York.[117] Even in those years, most of the guano was transshipped to Baltimore for distribution. Baltimore possessed two advantages. It was surrounded by a large area of poor soil where the development of improved agricultural systems had awakened farmers to the need for fertilizers that could not be supplied locally.[118] Secondly, Baltimore merchants had developed a flour trade with the west coast of South America. Ships in the flour trade found guano a convenient return cargo.[119]

Once established as the hub of guano distribution, Baltimore maintained that position. The principal agents and merchants generally were located there. They established channels of trade, communication, and finance between Baltimore and the smaller cities and towns of the Chesapeake region. Later, when the use of guano spread into the South, these channels were also extended. Generally, the consignees supplied the southern states via Baltimore, using local merchants as subagents.[120] These patterns had a large effect on the geography of the fertilizer industry later in the century. When other kinds of fertilizers were substituted for guano, they were often either imported into Baltimore or manufactured there and distributed from Baltimore through the mercantile channels established during the guano era.

The Effects of Guano on Agricultural Practices

Guano not only necessitated a new fertilizer distribution system, it also shaped the agronomic systems of many farmers. Although initially farmers considered guano a substitute for other materials that were in short supply, were too expensive or bulky, or had ceased to have their initial effects, guano soon became an essential fertilizer itself—especially in the worn-out soil areas surrounding Chesapeake Bay and to a lesser extent in similar areas of New Jersey and Long Island. In these areas, the availability of guano encouraged more intensive farming, allowed farmers to give up fallow rotations and the abandonment of worn-out land, and encouraged them to cultivate smaller areas more intensely instead of large areas extensively. Demand mushroomed as farmers discovered the miraculous effects of guano. Instead of being a substitute for other fertilizers, guano became the main fertilizer many farmers used.

As guano became their principal fertilizer, farmers were forced to modify their practices to accommodate it into their systems of farming. If guano had been broadcast on the fields months before planting, as had been the practice with lime, marl, ashes, and even stable manure, most of the valuable nitrogen in the guano would have volatilized and been wasted. If guano had been placed too close to seeds or plants, they would have been "burned" by the powerful fertilizer. Some of the first experimenters simply spread guano on top of their fields, but farmers quickly developed more efficient modes of application.[121] Farmers learned to apply guano near the plants, but not touching the seeds. They learned to place guano a few inches under the soil, instead of on top. By the late 1850s, a few farmers were even experimenting with drills to apply guano.[122] Rates of application also had to be worked out. Farmers experimented with from fifty to eight hundred pounds per acre, but eventually settled on two or three hundred as the most profitable rate.[123] Moreover, farmers learned to apply their guano in several dressings, spaced throughout the growing season for maximum efficiency.[124] All of these revised practices later facilitated the introduction of other concentrated fertilizers.

The most important changes were in the thinking of farmers. In the areas where guano was first introduced, farmers were already purchasing urban manures and soil supplements, but guano greatly reinforced and extended this commercial mentality. Thus, farmers became accustomed to purchasing large amounts of relatively expensive, concentrated fertilizers. Farmers also learned to accept fertilizers that acted rapidly and whose strength was used up in a year or two, as contrasted to lime, ashes, bones, or stable manure, whose effects were visible for much longer periods.[125] Most importantly, as a result of the use of guano, they came to expect that any other fertilizer they tried should be as highly concentrated.

Although guano had been substituted into the existing recycling system, its most important effect was to begin to break down that system and the recycling mentality. Farmers found it easier and cheaper to import guano than to recycle urban wastes. Even Solon Robinson, the tireless agricultural reformer and strong advocate of recycling New York's waste manures as fertilizers, realized that, compared with guano, city manure was a poor use of capital. He calculated that carting the equivalent in city manure of one load of guano thirty miles would require three hundred pairs of horses and three hundred fifty men. In his 1852 pamphlet on guano, he quoted several Long Island farmers who had come to the same conclusions.[126] Since guano eventually forged new geographical boundaries for the use of fertilizers as it pushed into areas (especially the South) where the old local and expanded recycling systems could never reach, the new mentality associated with guano spread to a far wider area than the old recycling mentality.

Instability of the System

The entire system of guano supply and use, which included the Peruvian government, consignees, charter shipping, exclusive agents, sundry dealers and merchants, agricultural experts, and the farmers who used guano, was effective, but at the same time it was unstable. Except for the Peruvian government, some of its favored officials, and its exclusive consignees, nearly everyone else in the system had incentives to seek changes. Shipowners wanted to capture some of the entrepreneurial profits from the guano trade and to avoid the irksome and costly regulations imposed by the Peruvian government. Dealers chafed under the heavy hand of the Peruvian government's exclusive agents. More than anyone, it was the farmers who were unhappy. They not only found the supply costly and undependable, but also began to experience problems with the extended use of guano as a fertilizer. The whole guano enterprise was laced with waste, corruption, and monopoly profits. That it could bear all of this is testimony to the marvelous fertilizing powers of guano. But problems made it inevitable that farmers, agricultural experts, and merchants would persist in seeking improvements in the supply system or alternatives to guano.

From the farmer's point of view, continuous price increases were the most burdensome problems. In 1845 Bartlett advertised guano for $44.80 per long ton (2,240 pounds) for large purchases.[127] Most farmers considered this cost excessive, especially when compared with lime and ashes, which generally cost about $4 or $5 per ton.[128] (Typical application rates of guano were only 200 to 300 pounds per acre instead of 6,000 pounds, so the actual cost per acre was only about $4 to $6 for guano instead of $12 to $15 for lime and ash. However, the effects of the latter fertilizers lasted five or six years, while the effects of guano were mainly exhausted in the first year.) As demand grew, the monopolists relentlessly increased the price. By 1847 the wholesale price was $50 per ton, by 1855 it was $55, by 1858 the price had reached $60. During the hard times that year, the consignees rolled the price back to $55, but the price was up to $62 by 1860.[129] By 1872 it had reached $90 for the best grades of guano.[130] Each time the price went up, farmers complained loudly and sought alternatives, but many had become so dependent on the product that they had little practical choice.

Farmers were also troubled by the long-term effects of using guano on their fields. They complained that guano produced an excess of straw, but not of grain; that guano exhausted the soil after repeated applications; and that the effects of guano were transitory, unlike more "permanent" manures such as lime, ashes, or stable manure.[131] These three related problems were caused by the imbalanced composition of guano. The percentage of nitrogen was too high and the percentages of phosphorus and potash too low. The excessive nitrogen stimulated rank growth of foliage, but not of grain. Moreover, excessive nitrogen tended to use up available phosphorus and potash in the soil until the land became "guano sick."[132] Then subsequent applications of guano did not have much effect and the soil appeared to be exhausted. The problem was especially severe in the Middle States where guano was often used alone.

Farmers using guano faced another problem that confronted the users of every commercial fertilizer introduced in the nineteenth century—the adulteration of the product. Unprincipled dealers easily adulterated guano by diluting it with large quantities of red earth, gypsum, charcoal, ashes, or a wide variety of other cheap and relatively worthless substances.[133] A member of New York's American Institute paid a surprise visit to a guano establishment and observed two sloop-loads of Long Island sand being unloaded and mixed with guano![134] Since farmers relied on their sense of smell to detect fraud, adulterators were careful to add enough real guano to give the mixture the characteristic pungent ammonia odor. Sometimes dishonest dealers purchased guano which had been damaged during transit by exposure to salt water.[135] This damp guano was available cheaply because its agricultural value was considerably diminished, but it was the ideal adulterant because the ammonia odor was much stronger than that of good guano. There was even a thriving market for secondhand guano bags (with the Peruvian government stamp) that some unscrupulous dealers refilled with spurious guano and sold as the genuine product.[136] Because fraud was so hard to detect without elaborate chemical tests farmers were not capable of performing, the uncertainties of using guano were needlessly increased. Probably many experiments, which otherwise might have been successful, failed because the guano used was impure.

Guano had proven its utility so clearly that farmers did not want to give it up. Instead they preferred to alter the supply system to lower the price and guarantee adequate quantities of high-quality guano. If alteration of the system was not possible, farmers wanted a substitute with the same fertilizing qualities.

Alterations to the System

Frustrated farmers blamed the monopoly exercised by the Peruvian government for ever higher prices and periodic shortages. In 1852 one critic calculated that if the Peruvian monopoly were eliminated, guano would cost only $27 instead of $55 per ton.[137] In 1853, Secretary of State William L. Marcy estimated that $30 to $35 would be a reasonable price.[138] A congressional committee investigating the situation in 1854 calculated that the "tax" American farmers paid to the Peruvian monopolists was at least $2,600,000 annually.[139]

As early as 1850, farmers in Maryland attacked the Peruvian government's monopoly.[140] The Maryland State Agricultural Society appointed a committee to investigate ways to remove some of the Peruvian government's restrictions on the guano trade.[141] The chairman of that committee, John Carroll Walsh, wrote to John M. Clayton, then secretary of state, to express the society's displeasure with the conditions of the guano trade.[142] Clayton already had taken steps to have John Randolph Clay, the American chargé d'affaires in Peru, suggest to the Peruvian government that it might benefit more from a moderate duty on the larger amounts of guano which would be exported if restrictions were removed than from the continuation of the present arrangements, but the Peruvian government was too dependent on the guano revenue and the services of the consignees to be willing to change the system.[143]

President Millard Fillmore stated in his annual message to Congress in 1850 that "Peruvian guano has become so desirable an article to the agricultural interest of the United States that it is the duty of the Government to employ all means properly in its power for the purpose of causing that article to be imported into the country at a reasonable price."[144] He promised that nothing would be omitted toward accomplishing that end. Throughout the next decade farmers continued to remind the government of this promise to improve conditions in the guano trade. In 1853, the Virginia Agriculture Society appointed agents to visit the consignees and President Franklin Pierce to discuss what could be done about the monopoly, but diplomatic efforts failed to end the monopoly.[145]

In 1854, the Maryland General Assembly passed a resolution calling on the federal government to make arrangements to end the guano monopoly, and two thousand Delaware citizens signed a memorial to Congress complaining of the exorbitant price and requesting that the United States convince Peru either to cede one of its guano islands or allow American vessels to take away guano without restrictions.[146] The pragmatic emissary, Clay, responded to dispatches from Washington by pointing out that there were great difficulties even in getting the Peruvian government to talk about the subject because so many ministers and other influential individuals were personally interested in the existing arrangements.[147] Clay did suggest, however, that Peru might be willing to sell one of the islands for sixty to eighty million dollars.[148] The only fruits of these petitions and memorials were the introduction of a bill regulating the trade between the United States and Peru, and the printing of twenty thousand extra copies of the congressional report on the guano trade.[149]

Concern over the monopoly price of guano did not abate. In 1855 *De Bow's Review* supported a proposed tariff designed to reduce the price of guano.[150] In March 1856 a protest meeting suggested by the Kent County Agricultural Society was held in Wilmington, Delaware.[151] Delegates from several of the mid-Atlantic states attended, but accomplished little more than to authorize another letter to the secretary of state and issue a call for a convention in Washington to discuss the problem further. That convention met on June 10 at the Smithsonian Institution, with delegates mainly from Virginia, Maryland, and Delaware.[152] The only product of this convention was a committee appointed to visit President Pierce and other government officials. Still nothing was accomplished. Complaints continued right up to the Civil War, but little could be done.

Farmers were not the only ones to complain about the Peruvian control of the guano trade. Shipowners also suffered from the high-handed ways of the Peruvian government. The chaotic, lawless conditions at the islands and the needless regulations caused them considerable difficulties and expenses. Captains and owners of ships occasionally protested, but little came of those protests.[153] Moreover, they felt that they could make larger profits if they could purchase guano freely at the islands and sell it on their own account instead of only chartering their ships to the agents of the Peruvian government.

Attempts to resolve the adulteration problem met with almost as little success as had the attempts to break the Peruvian monopoly. The most logical solution from the farmer's point of view was some form of regulation. States inspected or regulated many other products, including lime.[154] Therefore it seemed reasonable for states to provide a system of inspectors for guano. Pressure for such a system was greatest in Maryland, where the legislature passed a rudimentary but ineffective law in 1847.[155] During the next few years, farmers frequently demanded a better system of state inspections.[156] A stronger law, introduced in 1852 and finally passed in 1854, provided for a system of state inspectors to grade and stamp all incoming cargoes.[157] The Maryland inspection law by no means eliminated the potential for fraud, even for that state, since unscrupulous dealers quickly found ways to evade inspection or adulterate guano after it had been inspected.

Impact of Guano

Guano was the father or grandfather of the modern fertilizer industry. It helped transform an agricultural system and mentality based on the recycling of waste into a new system and mentality based on concentrated commercial fertilizers. As James F. W. Johnston, the noted Scottish agricultural chemist, had predicted in 1846, guano prepared farmers to accept other imported manures and concentrated artificial fertilizers which scientists might develop.[158] The urban-centered mercantile networks formed to accommodate the guano trade helped shape the geographical patterns of fertilizer production and distribution for most of the rest of the century. When the dramatic success of guano was followed by problems associated with its price, supply, adulteration, and composition, farmers and entrepreneurs quickly began a search for substitutes which could duplicate Peruvian guano's characteristics and effects. Eventually this search led, directly or indirectly, to the production of guano substitutes from fish, dried blood, and processed slaughterhouse wastes, and to the manufacture of superphosphates. These developments were the basis of the fertilizer manufacturing industry which emerged by the 1870s.

The important role of guano was mirrored linguistically by the persistence of the term.[159] Companies such as the Pacific Guano Company kept the word in their name even though their product contained no guano. Manufacturers called a wide variety of fish fertilizers, superphosphates, and other artificial fertilizers "guanos" to emphasize their supposed similarity to Peruvian guano. Farmers often referred to all fertilizers as "guanos," and the verb "to guano" became a widely used synonym for "to fertilize," especially in large areas of the South where the impact of guano was ultimately greatest.

CHAPTER IV

The Guano Island Mania

The dramatic success of Peruvian guano in the late 1840s and early 1850s and the Peruvian government's tight monopoly over the trade led to a search for ways to break the monopoly or find substitute sources of fertilizers. For some of the shipowners and merchants who dealt with the consignees of the Peruvian government, one alternative was to seek other island guano deposits. As these entrepreneurs avariciously sought such islands, they created what can best be called the "guano island mania."

The search for this "white gold" sent American explorers and fortune-seekers into most of the far-flung regions where guano islands might have been located. In some of its speculative excesses and high anticipations, this search resembled the rush for California gold that began the decade and the rush for "black gold" that ended it. Hundreds of guano islands were "discovered" and several dozen were legally attached to the United States to become America's first overseas possessions.[1] Attempts to claim and exploit these islands involved the United States in numerous protracted diplomatic disputes and led to a few minor military confrontations. The guano from these islands made an important contribution to the development of the fertilizer industry in the United States.

Ichaboe

Guano was so dramatically successful that almost as soon as the first cargoes from Peru arrived at the docks in Liverpool, farmers, agricultural experts, and entrepreneurs in Great Britain and the United States began to speculate about other possible guano deposits. British agriculturists considered the Hebrides and other Scottish islands as likely sources and even attempted to gather "guano" from the Flanborough cliffs in Yorkshire.[2] Practical experiments and chemical analyses, however,

54

quickly showed that these deposits were almost worthless as fertilizers.[3] An American agricultural journal suggested in 1843 that the flamingos, pelicans, and other birds on the Florida Keys might have produced deposits of bird excrements there similar to those found on the Peruvian islands.[4] Other Americans thought that guano might be found on islands known to harbor large colonies of birds in the Gulf of Saint Lawrence or off the coasts of Labrador and Newfoundland.[5] At an early 1844 meeting of the Farmers' Club of the American Institute in New York, one member reported having seen mounds of guano on some Florida islands and another suggested that all the islands of the East Coast should be examined for guano deposits.[6] None of these islands, however, had significant guano deposits, since the heavy rains quickly washed away the valuable parts of whatever bird excrement may have been deposited.

The first significant alternative deposit of guano was discovered in 1843 by a British entrepreneur who happened to read in the *Narrative of Four Voyages* by Benjamin Morrell, an American whaling captain, a description of an island off the southwest coast of Africa whose surface "was covered with bird's manure to a depth of twenty-five feet." This island sounded remarkably like descriptions of the Chincha Islands with their great flocks of seabirds and deposits of guano up to two hundred feet thick. The Liverpool merchant immediately sent out ships to locate the island, beginning an all-out assault by British guano merchants on Ichaboe, a tiny island located off the present-day Southwest African town of Lüderitz. By early 1845 as many as 450 ships were jockeying for position in the treacherous seas off Ichaboe attempting to load guano while thousands of men swarmed over the island digging the valuable manure. All of the frenetic activity on Ichaboe, however, quickly exhausted the island's meager deposits. When the last rock was scraped bare in May 1845, 300,000 tons of guano had been carried off.[7]

The first guano commercially imported into the United States may have come from Ichaboe. In September 1844 J. M. Thorburn of New York advertised a few hundredweight of this "superior guano" he had recently imported from Liverpool.[8] The following year E. K. Collins, the New York shipping magnate and a prime competitor of Cornelius Vanderbilt, imported an entire cargo of Ichaboe guano, which he apparently sold with some difficulty.[9]

While the amount of Ichaboe guano imported was small, it gave some farmers near New York City their first opportunity to try guano and accounts of its discovery may have inspired American adventurers to seek similar island treasures elsewhere.[10] Moreover, the sudden influx of Ichaboe guano into Great Britain temporarily glutted markets there and encouraged Gibbs and Sons to divert cargoes of Peruvian guano to the United States.[11]

The Lobos Affair

For several years after the Ichaboe episode, merchants brought cargoes of guano from Bolivia, Chile, Patagonia, and Saldanha Bay (near Cape Town) into Great Britain and occasionally the United States.[12] In addition, they attempted to introduce guano from other islands along the southwest coast of Africa.[13] With the exception of the Bolivian guano, which may have been the same as Peruvian guano, the other deposits generally proved inferior. By 1851 the London *Farmer's Magazine* concluded that alternative sources of guano as good as the Peruvian were unlikely to be discovered because valuable deposits of guano could only be found on islands near large shoals of fish in climates where rains were unknown. According to the article, moreover, such islands must be far from the haunts of man (who might frighten the birds) and must at the same time be accessible to shipping. The *Farmer's Magazine* reasoned (correctly) that such islands probably did not exist except along the coast of Peru and Southwest Africa.[14]

The first American targets in the guano island mania were the Lobos Islands, located about forty miles off the coast of Peru, and four hundred miles north of the Chincha Islands. Because the climate on the Lobos Islands was slightly wetter, the guano was inferior to that from the Chincha Islands.[15] Nevertheless, a group of American speculators assumed that proximity to the Chinchas made the Lobos group the most likely alternate source. The assault on the Lobos Islands was organized by Alfred G. Benson, a New York merchant whose firm was engaged in the California trade.[16] Return cargoes were always a problem in that trade. Obtaining a charter from the Peruvian consignee to carry Chincha guano was a possibility, but since charters were executed at a fixed rate per ton, the shipowners could not capture any of the large entrepreneurial profits to be had.[17] Merchants could do far better by taking guano from the Lobos Islands (without paying the Peruvian monopolists a cent) and selling the guano on their own account.

The inspiration for this venture apparently came from accounts in the British press early in 1852 of attempts by British entrepreneurs to have their government claim the Lobos Islands. The foreign office refused to cooperate, but the merchants arranged in 1852 for an expedition of twenty-five to thirty ships to the islands. The project collapsed when the Peruvian government made known its intention to protect its own claims to sovereignty over the islands, and when the British government refused to provide naval protection for the expedition.[18]

Alfred G. Benson apparently saw the reports in the British press and decided that the United States could exert a much stronger claim than Great Britain. He arranged for an associate, James C. Jewett, to write a letter on June 2, 1852, to Daniel Webster, then secretary of state, asking for information about the right to take guano from the Lobos Islands.[19] Webster replied only three days later that he considered it the "duty of this government to protect citizens of the United States who may visit the Lobos Islands for the purpose of obtaining guano."[20] At the same time, Webster arranged with the secretary of the navy to send a warship to the islands to protect American merchants there. It is unclear just why Webster made such a precipitate move to defend the interests of the American guano speculators in the Lobos Islands. Possibly, as William Cullen Bryant later charged, Webster and Benson were engaged in some sort of joint speculative venture.[21]

As soon as he learned that the secretary of the navy had given orders for naval protection, Benson proceeded to organize a large-scale speculative venture. He chartered "every vessel fitted for the work that could be procured . . . in the Atlantic" and instructed an agent to do the same in the Pacific. He later claimed to have chartered ninety-five ships with a capacity of 59,556 tons. He sent out from New York a ship loaded with "all the materials and laborers necessary for founding and sustaining for two years a colony on the Lobos Islands for the purpose of loading with guano such vessels as [he] should subsequently send there for that purpose."[22] To guard against failure, he arranged for a similarly equipped ship to be sent from Hawaii. The potential profits from such a venture were enormous. Benson could reasonably have expected to sell his guano on either the English or American market for at least $30 per ton and possibly much more. Since charter rates were about $17 per ton, and other expenses were low, Benson could have realized at least three-quarters of a million dollars if each of the chartered ships had made only one trip.

Despite extensive preparations, Benson's elaborate plan was foiled entirely. As soon as Don Juan de Osma, the Peruvian representative in Washington, learned of Webster's actions, he sent the secretary of state a strong letter of protest.[23] President Fillmore began to suspect that he had been misinformed, and the administration decided that "presumption of title" favored Peru. On August 16 the secretary of the navy suspended the June 16 order sending a naval vessel to the Lobos Islands, thus destroying Benson's hope of quick fortune and ending all American claims to Lobos guano.[24]

Two years later, Julius de Brissot, a New Orleans ship captain, launched an even more farfetched scheme to exploit the "considerable deposits" of guano on the Galápagos Islands. De Brissot, who was acting for Gen. José Villamil, the Equadorian chargé in Washington who had earlier attempted to colonize the Galápagos Islands, engaged Sen. Judah P. Benjamin of Louisiana to intercede on his behalf with the State Department and to travel to Equador to promote the scheme. The State Department undertook fruitless negotiations with Ecuador to secure American access to the alleged guano supplies and, if possible, to purchase the islands.[25] Unfortunately for de Brissot, no guano deposits existed on those islands, despite his early reports to the contrary.

Mexican and Colombian Guano

The inability of American merchants to obtain rights to take guano from the Chincha and Lobos islands forced them to look beyond Peruvian coastal waters for other islands with the same dry climate and multitudes of cormorants, gannets, and penguins. The first alternative deposits located by American merchants in 1852 or 1853 were on the Triangles and other small islands scattered about one hundred miles off the northern and western shores of Mexico's Yucatan peninsula.[26] Merchants considered these islands likely sources of guano because of their relatively dry climate and their large bird populations. The surface deposits on these islands, moreover, were soft, powdery, and generally similar in appearance to Peruvian guano. The location of the islands offered several significant advantages to American merchant adventurers. Since the islands were much closer to the East Coast markets, freight and insurance costs were much less than for the long trip around Cape Horn. Their location fit neatly into the well-established trading network between East Coast ports, especially Baltimore, and various Caribbean and Central American destinations. Finally, the islands themselves were so obscure and so far off the Mexican coast that American merchants, at least initially, could simply remove guano without having to submit to taxation or regulation by the Mexican government.

Although the Mexican guano appeared similar to the Peruvian variety, it was actually quite different in composition. The guano on the Mexican islands, and on most of the other islands subsequently discovered, had been deposited under climatic conditions less favorable than those along the Peruvian coast. Total rainfall, although still relatively light, was sufficient to leach out or decompose most of the nitrogenous material in the fresh bird excrement, leaving behind a deposit composed mainly of less soluble phosphates.[27] On the Mexican islands, some of the phosphates had also been leached out, leaving a product considerably less valuable than Peruvian guano. Nevertheless, farmers were willing to substitute this phosphatic guano for Peruvian guano, since the new variety cost less than half as much.[28]

By the spring of 1855, another type of guano, generally called "Colombian," appeared on the market. Despite its name, this guano came mainly from the approximately one hundred islands off the coast of Venezuela ranging from Los Monjes (The Monks) in the entrance of the Gulf of Venezuela, eastward through Aves, Los Roques, and Orchilla, to

Los Testigos.[29] The term, however, was loosely used to describe guano from such disparate locations as Roncador Cay near the coast of Nicaragua, another Aves Island about 120 miles west of Guadeloupe, and Sombrero Island in the midst of the British Virgin Islands.[30]

The guano from these islands was not only chemically different from the Peruvian variety (as the Mexican guano had been), but it was also physically different. Colombian guano was a hard, rocklike substance, hence its appellation "rock guano." Apparently, this physical difference delayed the introduction of Colombian guano into the American market for a few years, even though Colombian guano was actually more valuable since it contained much higher percentages of phosphates.

The introduction of Colombian guano set off an acrimonious debate among scientists regarding its origin and composition. First in print was an American chemist, Campbell Morfit, whose analysis of the hard guano lumps showed them to be "rich in phosphoric acid and lime."[31] Other chemists agreed about the high phosphoric acid content. What seemed to arouse the most interest in the scientific community, however, was the lumps' "mammillated exterior resembling enamel" which Morfit described.[32] Charles U. Shepard, the well-known Amherst mineralogist, who at the time also taught at the South Carolina Medical College and later was an important figure in the discovery of rock phosphate deposits in South Carolina, theorized in 1856 that this "petrified guano" was formed by some type of igneous activity.[33] Consequently, he classified three types of this guano as "pyroguanite minerals." Dr. A. A. Hayes, the Massachusetts state assayer and frequent contributor to the *American Journal of Science*, countered by suggesting that the rock-like crust had probably been formed by capillary action and evaporation.[34] This debate continued for the next several years in both the scientific journals and the agricultural press.[35]

Colombian guano was much more successful in the marketplace than the Mexican varieties. Although Colombian guano also lacked the high percentage of ammonia found in the Peruvian deposits, the islands on which it was found were "the richest known sources of large quantities of phosphoric acid," as even Yale's Samuel W. Johnson acknowledged.[36] Despite widespread prejudice against these phosphatic guanos, their demonstrated usefulness eventually helped shift farmers' ideas about what constituted useful fertilizers.

The Aves Affair

Inevitably, attempts by American merchants to exploit guano islands in the Caribbean drew the United States into international disputes. The first of these involved Aves Island, an isolated, uninhabited dot on the map approximately 350 miles north of the coast of Venezuela and 120 miles west of the Leeward Islands. To the Boston merchant Philo S. Shelton, who had previously imported Mexican guano, Aves must have seemed like an ideal source of guano. As its name implied, it was covered with an "immense quantity of birds." Moreover, its isolated position seemed to provide protection from the kind of governmental interference that American guano diggers sometimes experienced on Mexico's offshore islands. Therefore, in March 1854, Shelton and an associated Boston firm sent a ship under the command of Capt. Nathan Gibbs "on a cruise of discovery in the Caribbean Sea for derelict and desert guano keys or isles."[37] In April, Gibbs "discovered" Aves (although it was already well known), took possession of it for Shelton, and renamed it "Shelton's Isle." When Gibbs returned in July with men and equipment to set up permanent guano digging operations, he found that a rival Boston firm had already occupied the island. The two firms settled their differences peacefully by dividing the island, and proceeded to dig guano until that December when Venezuela sent a warship, abruptly expelled them from the island, and exerted her own tenuous claim to sovereignty.[38] After being expelled from Aves, the Boston operators filed claims for large damages against the Venezuelan government in a case which dragged on for years.

Ironically, the guano on the island was comparatively worthless as a manure.[39] Shelton may have been aware of the low value of his guano since he apparently sent two hundred tons of it to James J. Mapes's superphosphate plant in Newark, New Jersey, to be manufactured into "Chilian [*sic*] guano," which was presumably more valuable. To produce the "Chilian guano," Mapes simply ground Shelton's guano and added small amounts of Peruvian guano and ammonium sulfate![40]

The novel aspect of the Aves affair was that, unlike the entrepreneurs who surreptitiously worked other Caribbean guano islands, the Boston merchants attempted to claim Aves Island as United States territory. To buttress that assertion, they occupied the island, erected a "liberty pole," flew the Stars and Stripes, set up several houses, left some "American soil" (their ballast) on the island, and even transported some "American females" there.[41] Despite these elaborate precautions, Shelton's scheme failed. Elsewhere, other American entrepreneurs had suffered the same fate at the hands of foreign governments. Peru had blocked Benson's scheme for claiming the Lobos Islands, Venezuela had driven some Baltimore merchants from Los Monjes, and Mexico had expelled all foreigners gathering guano from the Alacranes off the coast of Yucatan.[42]

Shelton and his associates concluded that one of the principal reasons for the failure of his attempt to claim Aves was that it had not been made by a United States official or backed up by the United States government. Hence, with his April 19, 1856, memorial to Congress concerning his claims in the Aves case, Shelton submitted a bill drafted by his lawyer,

H. S. Sanford, that would have automatically extended United States sovereignty over any unclaimed guano island which was discovered and occupied by American citizens unless either the president or the Congress expressly declined to do so, and authorized the president to use the naval forces of the United States to protect any such claims.[43]

The Guano Island Act

Meanwhile, the rush for guano islands was accelerating in the Pacific as well as the Caribbean. Alfred G. Benson, the New York merchant who had been the prime mover in the Lobos affair, was setting up an even grander speculation in the mid-Pacific. This time he chose his targets and made his preparations more carefully. The area Benson had in mind, stretching from 150° to 180° west longitude and ranging 15° north and south of the equator, was not unexplored territory in 1855. Since 1820, American and British whalers had repeatedly visited this area.[44] J. N. Reynolds had interviewed whaling captains and assembled a list of islands in the area for the navy in 1835.[45] The United States Naval Expedition led by Capt. Charles Wilkes had crisscrossed the area charting a number of the islands between 1838 and 1842.[46] The whole area between Hawaii and Samoa was sometimes referred to as "American Polynesia."[47] If any nation could claim the islands in that area, it seemed reasonable that the United States could.

Benson first attempted to claim Baker Island (also called New Nantucket), located near the equator about 1,650 miles southwest of Hawaii, based on the supposed discovery of the island by an "ancient mariner," Michael Baker of South Dartmouth, Massachusetts.[48] Baker and another captain, Thomas D. Lucas of New Bedford, Massachusetts, who had sailed with him on some of his voyages, sold their rights through an intermediary in 1855 to the newly formed American Guano Company, of which A. G. Benson was president.[49] The following year, Benson assembled the documentation for a similar claim to Jarvis Island, several hundred miles to the east, again based on Baker's "discovery."[50]

Benson and his associates organized the American Guano Company, the legal shell for their enterprise, in New York City on September 1, 1855. The following month the company sent a delegation to Washington to present a memorial to President Franklin Pierce requesting that a warship be sent out "to inspect, examine, and survey . . . and take possession [of Baker Island] . . . under the flag of the United States, as the property of the [American Guano Company]."[51] Either the president was very much impressed with the facts of the case or Benson and his associates had considerable political influence. On October 20 the Navy Department issued orders to Comdr. William Mervine, commander of the Pacific Squadron, to dispatch to Baker Island at the earliest convenient opportunity one of the vessels under his command "with a view to ascertaining the correctness of the information, of protecting our citizens in their rights, and taking care of the interests of our country."[52] To further bolster its claim, the company also dispatched two ships, in December 1855 and January 1856, to explore Baker and Jarvis and to procure samples of the guano.[53]

Meanwhile, Benson and his associates prepared the statutory basis for their claims, with the help of Sen. William H. Seward of New York. Benson had already spoken before the Farmer's Club of the American Institute in New York about his undertaking. He may have been responsible for the resolution on the subject passed by the Club in 1855 that the United States should assert sovereignty over "all barren and uninhabitable" guano islands discovered by United States citizens and so far from any continent that no one could rightfully claim jurisdiction.[54]

Apparently in response to agitation and pressure from Shelton and Benson, in May 1856 Sen. William H. Seward of New York introduced a bill "to authorize protection to be given to citizens of the United States who may discover deposits of guano."[55] After this bill was reported out of committee on July 22, Seward took a leading part in the debate. Since his election to the Senate in 1847, Seward had been anxious to promote American commercial activity in the Pacific region.[56] Although he insisted that the "ragged rocks" on which the guano deposits had been found were "fit for no dominion" and that there would be no temptation to turn them into colonies once the deposits were exhausted, the bill still fitted his expansionist vision.[57]

The Senate finally passed the bill on July 24 after considerable debate over its expansionist tendencies and several minor amendments that attempted to balance the potential risks of the undertaking against the minimum incentives necessary to stimulate the enterprise while at the same time keeping the price as low as possible for agriculturalists. The House passed the bill without debate and President Pierce signed it on August 18, 1856.

The Rush for Pacific Islands

The Guano Island Act provided exactly the kind of incentives its supporters envisioned. It set off a mad scramble to locate guano islands, file claims, secure certification by the executive, and then profit from the protection and exclusive rights conferred under the act. By 1860 over fifty islands had already been claimed under the act.[58] A few of these were actually producing guano, but most had been claimed by speculators with the sole purpose of profiting by the sale of their exclusive rights.

The best placed of these speculators were Alfred G. Benson and his associates in the American Guano Company. Not only had they used their influence to assure congressional passage of the Guano Island Act, but they had also engineered the passage by the New York legislature of a general incorporation act for guano companies in March 1856.[59]

The company incorporated under this act and immediately proceeded to explore its Pacific possessions and physically establish its claims. Unfortunately for the company, after surveying the islands for the navy, Commodore Mervine had issued a strongly negative report concluding that it was impossible to land on this "'El Dorado' of the mercantile and agricultural interests" and that the supposed guano on the islands was worthless, since the islands were covered with vegetation, proving that the climate was not suitable for guano accumulation.[60]

Mervine's report did not reach Washington in time to influence debate on the Guano Island Act, but did cause the company considerable embarrassment. Fortunately, the two ships the company had sent to explore the islands were more successful. Agents of the company landed at Jarvis, formally took possession in the name of the company, built houses, left evidence of occupation, and obtained samples of the guano, although at the time the substance looked worthless to them too. Both ships also visited Baker Island although neither landed there.

To combat the negative publicity created by Mervine's report, Alfred G. Benson's half brother, George W. Benson, who had acted as the company's agent on Mervine's ship, wrote a long letter to the editor of the New York *Journal of Commerce* pointing out that Mervine had been prejudiced against the venture from the start and that as a "nautical gentleman" he was poorly qualified to judge the quality of a guano deposit, especially without physically examining the guano or even landing on the islands. Benson also noted that analyses of guano samples brought back by the other two ships had proved that Mervine's suppositions about Jarvis had been entirely wrong.[61]

To prove the correctness of its assertions, the American Guano Company in September 1856, requested that the president send a second naval expedition to survey the islands and bring back samples. The secretary of the navy dispatched a ship the following June. Comdr. Charles H. Davis of the USS *St. Mary's* landed on both Jarvis and Baker islands, formally took possession, surveyed them, and obtained samples of the guano. Davis, however, confirmed all of Commodore Mervine's skeptical findings. Davis, Mervine, and the captains of the American Guano Company's two ships all shared the popular belief that guano was worthless unless it came from a rainless region and smelled strongly of ammonia.[62] These negative opinions, however, had little effect, because an analysis by Joseph Henry at the Smithsonian of the guano samples that Davis had brought back proved them to be valuable.[63]

In April 1857 Alfred G. Benson's son, Arthur Benson, who had been sent out by the American Guano Company to set up operations, returned with five barrels of guano which he sent to the Patent Office to be distributed among farmers.[64] The following February C. H. Judd, acting as agent for the company, took twenty-three native Hawaiian laborers to Jarvis, erected buildings, laid moorings, and commenced commercial guano operations.[65] By late 1856 the company had provided the $100,000 bond, as required under the Guano Island Act, although the State Department did not issue the proclamation recognizing the company's rights until March 2, 1861.[66] The first commercial shipments arrived on the East Coast early in 1859.[67]

Digging guano on these islands was a difficult operation, especially on Baker where the anchorage was poor. Guano could only be loaded with great difficulty through pounding surf on ships moored precariously at buoys on the lee side of the island. Sometimes storms carried away the buoys and wrecked guano ships on the dangerous reefs near the islands. The guano itself lay in deposits ranging from one to six feet thick and often overlaid or underlaid with less valuable material easily confused with the phosphates. The digging and loading of the guano was a laborious task and living conditions were poor. All supplies, even water, had to be brought in by the company's schooner, which made the trip to Hawaii once every three months.[68]

Sometimes the American Guano Company imported guano itself, but usually it preferred to sell rights to the deposits to merchants such as William H. Webb, a leading New York shipbuilder and "solid citizen" who held a substantial interest in the Pacific Mail line.[69] In order to guarantee the integrity of the product, Webb sent out to the islands the first scientist associated with the guano business, James Duncan Hague, a young chemist who had been trained at the Lawrence Scientific School at Harvard, in Germany at the University of Goettingen, and at the Royal School of Mines in Freiberg. Hague later enjoyed a distinguished career as a mining engineer.[70]

Even before the American Guano Company's operations on Jarvis and Baker islands produced any guano, Alfred G. Benson began setting up still another venture involving Pacific islands. His initial targets were Howland, Christmas, Malden, and Arthur's islands, all in the general area of the other Pacific guano islands. As in the case of Jarvis and Baker, he found whaling captains who claimed to have "discovered" the islands or the guano on the islands.[71] Then he purchased rights to these islands and gave his son, Arthur Benson, who was employed by the American Guano Company to set up their operations on Jarvis and Baker, private instructions to land on and take possession of Howland Island in the name of "Alfred G. Benson and associates." Alfred Benson apparently planned to sell these claims to the American Guano Company at a handsome profit, but the company refused to purchase the islands it felt were rightfully its own.[72]

When the American Guano Company took steps to establish its own title to the islands, Benson broke with his former associates, withdrew from the company and on November 25, 1858, organized the United States Guano Company, which sent its first ship to Howland Island on November 12, 1858.[73] In the meantime, however, the American Guano Company had already sent its own agents to occupy the island. The resulting clash led to a long dispute that finally ended with the expulsion of the American Guano Company from Howland by force in 1862.[74]

Alfred G. Benson was not content with this four-island empire. In 1859 he purchased discoverer's claims to forty-two more islands or groups of islands, and he immediately sold to Webb and others the rights to remove guano. He filed bonds for these islands, and the State Department included them on the official list of guano islands.[75] At least a quarter of these islands, however, never existed. Several of the others almost certainly had no guano. Others were already inhabited, making it impossible to meet the requirements of the Guano Island Act that islands be "derelict."[76] As with the "discovery" of most of the guano islands in both the Caribbean and the Pacific,"exploration" seems to have involved nothing more arduous than poring over maps of the central Pacific Ocean in the comfort of a New York office.

The early exaggerated reports of the value of the guano deposits on the Pacific islands, coupled with the protections and inducements provided by Congress with the Guano Island Act of 1856, encouraged "fortune-seeking parties" to explore the Pacific in the hope of finding more guano islands.[77] The most important of these was C. A. Williams and Company of New London, Connecticut, whose expedition set out from Honolulu on January 19, 1859, to prospect for guano islands. On March 14 the company filed notice with the State Department of the "discovery" of Phoenix, McKean, and Enderbury islands in the Phoenix group and of Starbuck about a thousand miles eastward. Again, these were not true discoveries. All were well known, and both McKean and Enderbury had been visited by the Wilkes expedition in 1840. A month later, the company dispatched a ship from Honolulu with men and equipment, and began operations on Phoenix, McKean, and Enderbury under the name of the "Phoenix Guano Company" which marketed its product as "Phoenix guano." Like the American Guano Company, this company used Hawaiian laborers and supplied them quarterly from Honolulu. Initially, the company shipped actively, but Phoenix was worked out by 1871, and the other islands were abandoned by 1877.[78]

Several other Americans also claimed Pacific guano islands. One group of speculators, headed by a sea captain named William H. Parker and a lawyer named Richard F. Ryan, claimed six islands and formed a company to work their deposits. One of their islands, Johnston, was fully certified by the secretary of state on December 9, 1859, but only after Ryan had attempted to double-cross his partners and a conflicting claim by Hawaii had been settled.[79] In 1859, a United States naval officer claimed to have discovered guano on French Frigate Shoal in the Hawaiian Islands, and San Francisco interests claimed Lisianski, also in the Hawaiian chain.[80] Ironically, the American speculators overlooked the rich phosphate deposits on Ocean and Nauru islands only slightly further west, probably because these were not bird islands and their ancient phosphate deposits did not resemble guano in appearance.[81]

Americans did not have the Pacific Islands entirely to themselves. In 1859 the French claimed Clipperton Island, which is about seven hundred miles southwest of Mexico.[82] British and Australian interests were also active. They began working Malden and Starbuck in the 1860s, although the American Guano Company also claimed Malden. After the American companies had abandoned most of the islands, a British firm occupied and worked many of them in the 1880s.[83]

Caribbean Guano Islands

In 1860 Jefferson Davis criticized the Guano Island Act in a Senate debate as "extraordinary in its character, and . . . really dangerous in its effect." He complained that the act had already involved the United States in more diplomatic questions and could more seriously disturb the peace of the country "than any other law or any act which probably [had] occurred in the whole history of the nation."[84] Most of the disputes Davis referred to had occurred in the Caribbean, where the act had added impetus to the search for guano islands already under way. Because the guano islands in the Caribbean were generally near enough to the mainland or to one of the larger West Indies islands that some other nation could plausibly exert sovereignty, difficulties were bound to occur when American adventurers made claims under the Guano Island Act.

One of the first islands claimed under the act was Sombrero, a ninety-five acre rock located in the Anegada Passage in the Leeward Islands. Since Sombrero was surrounded by British islands, it seems unlikely that it could have qualified under the act as "not within the lawful jurisdiction of any other government." But John E. Gowan and Franklin Copeland, who had been expelled from Los Monjes by Venezuela late in 1855, claimed early in 1857 that one of their captains had discovered a deposit of guano on Sombrero the previous December. They hoped that the new law would make their tenure on the island more secure, but the State Department never officially recognized the claim. Several American companies worked the island for a few years and marketed their product as "Sombrero guano," but in 1868, without protest from the United States,

Great Britain formally took possession of it as one of her Virgin Islands.[85] British companies continued to work the island's guano deposits for several more years.[86]

Shortly after the "discovery" of the Sombrero deposits Joseph W. Fabens, a Democratic politician who had held minor diplomatic posts in Latin America, reported that he had discovered guano on the Swan Islands, located slightly over one hundred miles north of the Honduran coast. With Duff Green, the politically well-connected Washington journalist who had been involved in a wide variety of speculative ventures, and with Charles Stearns, Fabens formed the Atlantic and Pacific Guano Company. For a few years this company worked the Swan deposits as well as the Aves deposits. Then in 1862 the New York Guano Company purchased the Swan Islands.[87] Eventually the Pacific Guano Company of Boston and Woods Hole, Massachusetts, obtained the islands and worked them sporadically until the beginning of the twentieth century.[88]

Possibly the most important guano deposit claimed under the Guano Island Act was on Navassa, about thirty miles west of Haiti. Capt. Edward K. Cooper of Baltimore claimed that he or an associate had discovered the deposit and began guano operations on Navassa in 1858.[89] At first the Philadelphia Guano Company acted as agent for at least some of the guano, but in 1864 Cooper organized the Navassa Phosphate Company, which operated the island for most of the rest of the century.[90] Like most of the guano islands, Navassa was not a pleasant place to work. In the mid-1860s, there were about thirty white officers and mechanics on the island and about one hundred eighty black laborers.[91] Later, because of the difficulty of attracting labor to the desolate island, Cooper contracted with the state of Maryland for prisoners from the state penitentiary. Banishment to the island was regarded as "the ultimate of earthly punishment." In spite of the "hatred, reckless rebellion, bloodshed, death, and hurricanes which characterized the whole scheme," guano was shipped in large quantities, and Cooper made a fortune from it.[92]

Cooper's operation on Navassa was the cause of the first diplomatic scrape after passage of the Guano Island Act. In 1858, before his title had been certified, Haiti attempted to challenge Cooper's claim. The United States responded by sending a warship to visit Navassa and made a show of force at Port-au-Prince.[93] To solidify his claim, Cooper wanted Congress to pass special legislation recognizing his title.[94] Virginia's Sen. James Mason introduced a bill to accomplish this, but since Haiti had already abandoned its claim, the special legislation appeared unnecessary, and the bill died in committee.

American interests claimed several other islands scattered in forgotten corners of the Caribbean before the guano island mania died. In November 1858 Capt. Richard Daulby, employed by the Baltimore merchant W. T. Kendall, sailed from Baltimore "to see if he . . . could not discover a place where he could get some guano."[95] He quickly discovered a deposit on Cay Verde in the Bahamas, but only had time to make two trips to remove guano before the British expelled his men and took over the island in March 1860.[96] The United States acquiesced in this expulsion without protest, as it did the same year when the Spanish expelled American guano diggers from Mona and Monita islands in the Mona Passage between Puerto Rico and the Dominican Republic.[97]

After his men were expelled from Cay Verde, Kendall and a competing Baltimore firm attempted to claim Alto Vela, lying about twenty miles off the southern coast of the Dominican Republic and only eight miles from Beata Island, which was clearly Dominican territory, but the Americans were quickly expelled by the Dominicans.[98]

During the 1850s adventurers discovered a number of other guano deposits in various parts of the world. In 1853 the British press reported discoveries of guano in Shark Bay on the west coast of Australia and in the East Indies.[99] A few years later a British captain claimed to have found guano deposits on the Kuria Muria Islands five miles off the southern coast of the Arabian Peninsula, but a commercial expedition to the islands failed to find any significant deposits.[100] By the end of the decade American entrepreneurs were advertising "California" guano from islands along the shores of Baja California. Although the California guano was moderately good in quality, the amounts available were limited. American interests never tried to claim the islands involved, probably because most of them were too close to the Mexican mainland for any questions about their ownership.[101]

The guano island mania subsided rapidly after the 1850s. Entrepreneurs discovered and worked guano deposits on Aruba, Curaçao, St. Martin, Redona, Pedro Keys, and Orchilla in the Caribbean. Americans never claimed any of these islands, and with the exception of Orchilla, most of the guano on them apparently went to Great Britain.[102] Americans did claim about two dozen more Caribbean islands after the Civil War. Some of these were islands off the coast of the Yucatan which had long been the source of "Mexican guano." Others were scattered along the coasts of Honduras, Nicaragua, and Jamaica. Most of these islands, however, produced little or no guano.[103] In addition, Americans attempted to claim several islands scattered between Hawaii and Japan during the last decades of the century.[104]

Patterns of the Mania

During the guano island mania, the pattern of "discovery" and exploitation was determined by three factors. First, the entrepreneurs depended heavily on information gathered as a result of whaling voyages, mercantile activity, and systematic exploration. Second, previously existing trading patterns and the need for return cargoes directed the search in the Caribbean and Pacific. Third, as in most aspects of the development of the fertilizer industry, the course of discovery and exploitation advanced by the process of marginal substitution. When initial efforts failed to dislodge the Peruvian government's system of monopoly and exclusive consignees, American entrepreneurs first attempted to claim islands as near as possible to the original Peruvian deposits. Then, when Benson's Lobos venture failed, merchants sought islands elsewhere with the same dry climate and immense bird populations. In this search, merchants first exploited deposits such as those found on Mexico's offshore islands, which looked like Peruvian guano even though they were chemically different. It took the merchants longer to exploit the Colombian guano, because it did not look the same as the Peruvian, although it was actually a more valuable fertilizer than the Mexican variety. No one discovered the immense deposits of rich phosphate rock on Ocean and Nauru islands until the end of the century because the islands shared none of the physical characteristics of the guano islands and the phosphatic deposits did not look like guano.

Wherever the search for guano led, the entrepreneurs involved were characterized by an "exploitation mentality." They quickly moved in whenever guano deposits were reported, and began digging up whatever guano they could find. Sometimes they removed virtually entire islands. Frequently they exhausted rapidly deposits that had been accumulating for thousands of years. Entrepreneurs were equally willing to exploit whatever human resources they could find. Although conditions were seldom as bad as on the Chincha Islands, most of the guano islands were bleak, isolated, undesirable places to live. Entrepreneurs were forced to use prisoners, sailors, or whatever natives they could coerce or entice from nearby islands.

Most of the entrepreneurs felt little compunction about obeying the laws of nations the islands might rightfully belong to. If they could exploit a guano deposit before the owner discovered it, they felt perfectly free to do so. The entrepreneurs often felt no more compunction about being honest with the farmers purchasing the guano. They frequently misled purchasers about both the origin and the composition of the guano they were selling. Most of the men involved came out of merchant backgrounds, but apparently the high risks and dubious respectability of the guano island enterprise attracted the more freewheeling members of the merchant fraternity.

Results of the Mania

The guano island rush left behind a geographical and political legacy. Roy Nichols, who has thoroughly examined the diplomatic aspects of the episode, believes the entrepreneurs who claimed the guano islands were "advance agents of American destiny."[105] In a sense they were, although the islands themselves were small, uninhabitable, and never of any strategic importance.

The amounts of phosphatic guano brought into the United States were never large. Between 1853 and 1873, with the exception of the Civil War years when imports were lower, the annual imports averaged about fifteen thousand tons per year. Almost all of this came from the Caribbean, except for the period between 1857 and 1862 when about two thousand tons per year came from the Pacific islands. In the 1880s importations reached forty thousand tons a year, and then dropped off rapidly until the last recorded shipment in 1900. After 1870, when the treasury began tabulating separate statistics for the bonded United States guano islands, they usually accounted for about half of the imports. Most of this came from Swan and Navassa. The amount of phosphatic guano imported was much less than the amounts of Peruvian, which averaged almost sixty thousand tons annually in the decade before the Civil War temporarily cut off the trade.[106]

The brief adventure with the guano islands never exerted a very significant impact on total American agricultural production. Even in the peak years there was only enough phosphatic guano imported to fertilize about one out of every five hundred acres of tilled land at minimum rates of application. These figures ate deceptive because the use of these guanos was concentrated in areas in the East, especially in Maryland, Delaware, and Virginia; but even if all the phosphatic guano had been used in those three states, there would only have been enough for about one out of every fifty acres cultivated. The significance of the phosphatic guano trade, however, was far out of proportion to its volume because it allowed numerous entrepreneurs to enter the guano business who might not have been able to do so otherwise and because the chemical composition of the phosphatic guanos forced farmers to reevaluate what constituted a valuable fertilizer and encouraged entrepreneurs to undertake processing operations.

CHAPTER V

The Impact
of Phosphatic Guano

As entrepreneurs attempted to find a substitute for Peruvian guano, they sought to locate other sources for the same product or produce a substitute which could be sold and used like Peruvian guano. Inevitably, however, these attempts introduced inadvertent changes. In the case of the guano islands, even though islands with similar climates and fauna were sought, the guano on them differed significantly from that of the Chincha Islands. These differences encouraged farmers to accept new theories of fertilization and entrepreneurs to undertake further processing to make the guanos effective fertilizers. In the process, farmers were weaned away from their age-old reliance on "organic" fertilizers and entrepreneurs were nudged into processing nonorganic raw materials into fertilizers. Thus, even though the guano island mania was short-lived and the total amount of guano imported from these islands was small, it did have a major impact on agricultural practice and on the shape of the emerging agricultural chemical industry.

Phosphatic Guanos and the Mineral Theory

The entrepreneurs importing Mexican guano in the early 1850s immediately faced the problem that confronted all subsequent attempts to utilize the non-Peruvian guanos. Most farmers and chemists then estimated the value of a fertilizer primarily by its "ammoniatical" content. Mexican guano was accumulated bird excrement like the Peruvian, but it had been exposed to enough moisture to destroy most of the nitrogenous material. In the early 1850s the prevailing opinion was that the remaining phosphates and "effete organic matter" were much less valuable than Peruvian guano.[1] One respected English chemist concluded that Mexican

71

guano was of a "very low character . . . not worth importing."[2] Wide variations in the samples tested by chemists made the product look even less promising.

Because the phosphatic guanos were not chemically equivalent to the Peruvian variety, merchants selling phosphatic guano could not appeal to the same scientific theories to promote their product as had the importers of Peruvian guano. In the 1850s two scientific theories competed to explain how plants fed. One, known as the "ammoniacal theory," maintained that the most important substance for the growth of plants was ammonia obtained either from the atmosphere or from organic material in the soil. The British chemists John Bennet Lawes and J. H. Gilbert were the principal proponents of this theory, which adequately explained the effects of Peruvian guano. Opposed to this was the "mineral theory" that phosphates and other minerals in the soil were more important than ammonia in plant development. Justus Liebig, the renowned German agricultural chemist, was the principal proponent of this theory, first introduced in the 1843 version of his *Animal Chemistry*. In this country Leonard D. Gale, the chief chemical examiner at the Patent Office from 1846 to 1857, was the mineral theory's strongest supporter.[3]

The importers of the phosphatic guanos from the Pacific and Caribbean seized upon this mineral theory as ideally suited to promote their product. Beginning with the introduction of the Mexican and Colombian guanos in 1854 and 1855, advertisements provided clear explanations of the plant's need for phosphates and explained the necessity of phosphatic fertilizers, especially for the "worn-out lands" of Maryland and Virginia. To distinguish their product from Peruvian guano, the promoters claimed repeatedly that "it is the *mineral* and not the *ammoniacal* properties of guano, which give it value as a manure."[4] Furthermore, they attempted to show that the ratio of nitrogen to phosphates in Peruvian guano was so high that it would both overstimulate plants and exhaust the soil by using up the natural reserve of minerals.[5]

Promoters of the phosphatic guanos relied extensively on Liebig's theories to support this type of assertion. A pamphlet published by the American Guano Company in 1857 quoted from Andrew Ure's *Dictionary of Arts, Manufactures and Mines*, which cited Liebig, to show that "of all the principles furnished to plants by the soil, the phosphates are . . . the most important."[6] The pamphlet, however, conspicuously ignored a prominent assertion on the previous page of Ure's *Dictionary* that guano from which the ammonia had been washed out was worth considerably less than guano with a high percent of ammonia remaining. Two years later, the United States Guano Company made an even bolder assertion.

It has thus been demonstrated by the unbiased and most elaborate investigations of scientific men who have written solely for the benefit of agricultural interests, that the popular belief in the value of ammonia as a fertilizer, is utterly erroneous, and that its use does not *enrich*, but impoverishes the soil; and the Trustees deem it fortunate for the world that a Liebig and a Gale (not to mention others) have been able to expose so mischievous an error, by demonstrating that it is nutrition, not stimulus which restores an exhausted soil to its original fertility.[7]

To buttress these claims, the company quoted extensively from Liebig and from a chemistry text by his English student and associate, James Sheridan Muspratt, to which Harvard's pioneer agricultural chemist, Eben N. Horsford, had added a supplement.

Other promoters also cited Liebig to support their product. Wood & Grant's pamphlet introducing their phosphatic guano from Sombrero had extracts from Liebig's *Familiar Letters on Chemistry* on the cover and contained five pages of quotes from his *Agricultural Chemistry*.[8] The Philadelphia Guano Company also quoted heavily from Liebig in advertisements for its Sombrero, Colombian, and Mexican guanos to prove "the *necessity* of phosphates" for the growth and maturation of crops.[9] William Webb's 1862 pamphlet advertising Baker and Jarvis Island guanos quoted from Liebig and Muspratt and reproduced an 1860 report in which Liebig highly recommended Baker and Jarvis guanos because of their high percentage of phosphates.[10]

The promoters of the phosphatic guanos attempted to translate these scientific theories into terms farmers could readily understand. One method was to equate "stimulus" with ammonia and to equate "nutrition" with phosphates. The United States Guano Company *Report* concluded its arguments in favor of phosphatic guanos over ammoniacal fertilizers with this analogy, which must have pleased temperance advocates:

> It is nutrition, not stimulus, which restores an exhausted soil to its original fertility, and that to drug the ground with ammoniacal manures, and expect good crops, is not a greater fallacy than to feed infants with alcoholic stimulus and expect to raise a sturdy generation of men.[11]

The promoters of phosphatic guanos also attempted to explain the effectiveness of their products by comparing them with bones. American farmers had been using bones, at least to a limited extent, for several decades before the introduction of phosphatic guanos. Even farmers who had not used bones were aware of their utility as a fertilizer. Advertisements frequently compared bones with phosphatic guanos and pointedly referred to calcium phosphate, the principal fertilizing ingredient in both bones and phosphatic guano, as "bone phosphate of lime."[12] Often the comparisons with bones were even more explicit, such as one promotional pamphlet for the Navassa guano that included an elaborate description of the long-lasting fertilizing effects on the Waterloo battlefield of the bones of fallen soldiers.[13]

Despite these various explanations of the usefulness of the phosphatic guanos, many agriculturalists remained unconvinced. The *American Agriculturist* refused to carry advertisements for Colombian guanos when they were first introduced. Later it carried advertisements for the Pacific guanos only reluctantly, since its editors felt that any fertilizer composed principally of phosphates and other mineral elements was not valuable.[14] The editors of the *Agriculturist* were not prepared to accept Samuel W. Johnson's recommendation of some phosphatic guanos on the basis of scientific evidence, even though they admired most of Johnson's work. In a generally laudatory review in 1859 of Johnson's *Essays on Peat, Muck, and Commercial Manures*, the *Agriculturist* pointedly criticized the value Johnson placed on phosphatic guanos, stating "that in this instance practice will not sustain the apparent indication of science."[15]

The following year the *Agriculturist* reiterated its firm conviction that "purely mineral or phosphatic guanos . . . will not prove profitable or satisfactory.[16] The *Agriculturist* echoed the thoughts of many farmers when it maintained that "the value of any concentrated manure depends mainly on the amount of organic matter it contains."[17] For instance, in 1862 a committee of farmers reported to the American Institute that the Sombrero guano imported as a substitute for bones was more suitable for dock building.[18] Some of the reluctance to accept the phosphatic guanos may have been due to the attachment of farmers to well-established theories. At least some resistance, however, was due to the wide variety of analytical results scientists reported, the variability in the quality of the phosphatic guano imported, the inexplicable failure of the guano on some soils, and the misleading information in some of the advertisements.[19]

Ironically, even though the speculators involved in the various phosphatic guano enterprises may have selectively chosen scientific theories and stretched the truth on occasion for their own benefit, they were correct in asserting that phosphates were essential for the development of plants. The speculators had not intended to introduce any new scientific theories. In fact, their original goal was to find guanos with the same organic matter as the Peruvian variety. The high phosphorus content and low nitrogen content in the guanos they discovered, however, forced them to expound the new mineral theory. Their propaganda, combined with the evident success of these guanos in many fields, helped shift farmers away from the old ammoniacal or "organic" theory toward the acceptance of a more modern theory which recognized the importance of minerals such as phosphates. This successful transition toward the presently accepted theories took place despite the many weaknesses and inconsistencies in Liebig's work that those promoting the phosphatic guanos conveniently overlooked.

Since the phosphatic guano generally lacked the strong odor of ammonia which farmers expected, the success of these guanos may also have helped change the age-old prejudice of farmers that "a good manure must have a bad smell." Samuel W. Johnson pointed out in 1859 that it was difficult to convince farmers that a phosphatic guano with "no more smell than sand" was valuable.[20] Once the efficacy of one odorless fertilizer had been demonstrated, however, it was much easier to get farmers to accept similar materials, such as superphosphates manufactured from phosphate rock. The success of these guanos, moreover, helped establish the usefulness of phosphates as a fertilizer and thus led to the practical acceptance by farmers of the necessity of providing not only "organic" (i.e., nitrogenous) but also "inorganic" (i.e., mineral) fertilizers for their crops.

Manipulation and Processing of Guano

Although merchants advertised and sold guanos from the Caribbean and Pacific for direct application to the soil like Peruvian guano, the use of unprocessed phosphatic guanos was limited. Many practical agriculturalists had resolved the dispute between the ammoniacal and mineral theories by deciding (correctly) that both were right and that the ideal fertilizer should contain both ammonia and phosphates. Therefore, almost from the introduction of the Mexican and Colombian guanos, recommendations appeared suggesting that the material be combined with Peruvian guano to provide a mixture more effective than the Peruvian guano alone, less likely to exhaust the soil, and significantly cheaper.[21]

Entrepreneurs quickly attempted to capitalize on these recommendations by producing "manipulated guanos." These were simply premixed combinations of Peruvian and phosphatic guanos with additions that depended on the integrity or deviousness of the manipulator. By 1857 Baltimore and New York entrepreneurs offered numerous products such as "Kettelwell's Manipulated Guano," "Reese's Original Manipulated Guano," and "De Burg's No. 1 Manipulated Guano." Generally the manufacturers claimed that their mechanical equipment produced a more uniform product than the farmer could mix himself. In addition, they maintained that because of their location in the principal ports of entry, they had access to the best imported guanos, which they were able to buy more cheaply in bulk. Furthermore, these manipulators advertised that their guaranteed analyses, sealed bags, and personal integrity ensured that farmers would get a consistently reliable product.[22]

By 1860 at least five manipulators were operating in Baltimore, about the same number in Richmond, several in Petersburg and Alexandria, Virginia, and several more in New York and Connecticut.[23] Some of the producers claimed to mix Peruvian and phosphatic guanos in equal proportions, but it is likely that many used larger portions of the cheaper phosphatic guanos to cut costs.[24] For farmers, the most significant advantage of the manipulated guanos may have been that they were ground finely, thus greatly increasing the efficacy of both the Peruvian and the phosphatic guanos incorporated in the mixture. The manipulated guanos were also much closer to an ideal balanced fertilizer than either the Peruvian or the phosphatic guanos had been.[25] Because the ratio of nitrogen to phosphorus was lower, they did not tend to produce excess growth of straw or foliage, as Peruvian guano sometimes did. Because of the higher percent of phosphorus, the effects of the fertilizer would last over a longer period of years.

If the manipulator was honest, the farmer paid very little for the service. Generally the price of a manipulated guano was at most a dollar or two more per ton than would have been the cost if farmers had purchased the ingredients separately. As in so many aspects of the fertilizer business, however, not all dealers were honest.[26] During the manufacturing process manipulators could easily introduce adulterants farmers could not readily detect. The prevalence of such adulterations or farmers' fear of such adulterations may have slowed the acceptance of these manipulated guanos.

While manipulated guanos were better than either Peruvian or phosphatic guanos alone, they were still not the ideal fertilizer. The principal problem was that the phosphates in most of the guanos from the Caribbean and Pacific were relatively insoluble. The more precipitation that fell on the guano islands, the less likely that any soluble phosphates would remain in the guano. Although some chemists initally maintained that some of the phosphatic guanos were "natural superphosphates," farmers and manufacturers gradually realized that phosphatic guanos would work more rapidly if they were made more soluble.[27]

In the early 1840s the English gentleman farmer and agricultural experimenter John Bennet Lawes had pioneered the manufacture of soluble phosphates by treating bone, bone ash, and mineral phosphates with sulfuric acid to produce a "superphosphate." (See chapter 7 for a fuller description of this development.) In the early 1850s several manufacturers around New York and Baltimore also began producing superphosphates, using bone or bone charcoal as the phosphatic raw material. Since chemists clearly understood the similarity between bones and phosphatic guanos, it was a natural step to substitute phosphatic guanos for bone in the manufacture of superphosphates. By the later 1850s several Baltimore merchants had already undertaken the manufacture of superphosphates from phosphatic guano and more followed in the early 1860s. The use of phosphatic guano as a raw material for superphosphates solved two problems for the fertilizer industry. First, it overcame the limited supply of bones that had severely constrained the production of superphosphates. Second, this processing step created a ready market for the phosphatic guanos, thus producing profits for the merchants who imported those guanos.[28]

The Impact on Entrepreneurs

Many of the urban entrepreneurs who dominated the emerging agricultural chemical industry in the 1860s and 1870s began their careers either importing or processing phosphatic guanos. The Peruvian guano monopoly had confined American entrepreneurs to secondary roles, but the importation of phosphatic guanos opened up lucrative opportunities for entrepreneurial profits. Once established in this trade, the importer often found that the nature of the guano made further processing necessary to ensure a satisfactory or complete fertilizer. Therefore, he was likely to commence either manipulating or acidulating his guanos. Once a manufacturer had built the necessary plants, set up the requisite distribution system, and established a reputation for his product, it was very easy for him to substitute other raw materials, such as South Carolina phosphates when they became available, or to change his formulas as scientific or practical evidence dictated.

A number of Baltimore merchants followed this path to success in the fertilizer industry. They began as general merchants, gradually specialized in guano, then undertook guano manipulation, and finally emerged as leading fertilizer manufacturers or distributors. For instance, William Whitelock, who later became a leading figure in the industry, came from Norfolk, Virginia, where he had received mercantile training. In 1845 he established a small shipping business between Baltimore, the southern states, and the West Indies. He was one of the merchants who tried to sell the first commercial cargo of Peruvian guano brought into Baltimore in 1845. Later he began to import phosphatic guanos, which made a convenient return cargo in his trade with the West Indies. By 1854 he was advertising Mexican guano along with African and Peruvian guanos, and claimed to be "the oldest house engaged in the trade." By the 1860s he had turned his principal interest to the manufacture of fertilizers. His plants in Baltimore's Federal Hill area produced "Whitelock's Vegetator" and other fertilizers for the next thirty years.[29]

The career patterns of several other Baltimore merchants followed the same path. R. W. L. Rasin, another leader in the industry after the Civil War, first held positions with the Philadelphia Guano Company, which imported Venezuelan and other Caribbean guanos in the late 1850s. He later worked with the Sombrero Guano Company and then joined Capt. Edward K. Cooper, who operated Navassa Island. Eventually Rasin became general agent of the Navassa Phosphate Company. In 1872 Rasin and Captain Cooper formed a partnership in the Rasin Fertilizer Company, which in 1882 advertised that it had "the most extensive and permanent works in the United States."[30] Similarly, B. M. Rhodes began in Baltimore in 1854 as a "general commission merchant" who dealt in Mexican and Peruvian guanos.[31] He continued to import various phosphatic guanos for a number of years.[32] By 1857, however, he also started producing "Rhodes's Superphosphate of Lime," which became one of the most prominent brands of fertilizer in the following decade.[33]

Several druggists and chemists also followed the path from guano manipulation in the 1850s to fertilizer manufacture in the 1860s. Gustavus Ober, the founder of another important firm, was the son and grandson of Georgetown merchants. At fifteen he entered the drugstore business in Philadelphia, and at 21 he graduated from the College of Pharmacy in that city. He remained in the drug business there until 1840, when he moved to Baltimore and opened his own wholesale drug business. In 1856 he became general agent of his brother-in-law John Kettlewell, who was already manipulating guanos. He manipulated guano with Kettlewell until the Civil War cut off their markets in the South and Kettlewell died. After the war Ober resumed the manufacture of fertilizers, and in 1869 formed the firm of G. Ober & Sons, which became one of the principal superphosphate manufacturing firms in Baltimore.[34] Similarly, Richard J. Baker began his career in the drug and dyestuffs business, then manipulated guanos, and later entered the fertilizer manufacturing business, eventually becoming president of the Piedmont Guano and Manufacturing Company.[35]

The manipulation of phosphatic guano had an equal impact on the career of John S. Reese, who began as an "apothecary and druggist," but by 1853 turned to manipulating guanos in partnership with John Kettlewell. After the Partnership broke up in 1857, he continued to manufacture manipulated guanos under his own name. The *American Farmer* specifically commended him for his honesty in revealing the constituents of his fertilizer. He did not, however, undertake the manufacture of superphosphates, and by the early 1860s he withdrew entirely from manufacturing. For the rest of his career he specialized in fertilizer distribution, after 1864 as general agent for the southern states for the Pacific Guano Company of Woods Hole, Massachusetts.[36]

The phosphatic guano trade helped solidify Baltimore's position as the center of the emerging fertilizer industry. Because of the demand for fertilizers generated in the Chesapeake region, Baltimore had already become a leading distribution center and one of the two principal ports of entry for Peruvian guano. This facilitated the introduction of phosphatic guanos in the same area. Because of Baltimore's trading connections with Latin America, moreover, its merchants had an advantage over those in other East Coast cities in entering the phosphatic guano trade in the Caribbean, where most of the important deposits were found. Although Philadelphia, New York, and Boston merchants played a role in opening that trade, most of the merchants involved had Baltimore connections. Once the trade was well established in Baltimore, the merchants were forced to undertake manipulation or acidulation there. The development of the trade networks centered in Baltimore, the construction of plants and equipment in Baltimore, and the accumulation by that city's merchants of expertise in the field made it much easier for Baltimore to become the center for the manufacture of fertilizers from South Carolina rock phosphates in the 1870s.

The Pacific Guano Company

Perhaps the best example of the way phosphatic guanos provided a path into the fertilizer manufacturing industry was the Pacific Guano Company of Woods Hole, Massachusetts. This company grew out of the Boston mercantile firm of Glidden and Williams, which was interested in shipping to the Pacific Coast in the 1850s and later set up the California Packet Service. In the late 1850s the firm sought to solve the persistent problem of finding return cargoes from the West Coast by entering the guano trade.

Members of the firm established the Pacific Guano Company in 1861. The company later claimed to have purchased Howland Island in the Pacific, although it may only have purchased from Alfred G. Benson's United States Guano Company the right to remove guano. For several years the company imported and marketed this "Pacific Guano." Initially they dried their guano in the vats of a deserted saltworks on Spectable Island in Boston Harbor. Soon the company decided that their product could be made more similar to Peruvian guano by ammoniating it with fish scrap which entrepreneurs along the East Coast had been manufacturing from menhaden fish for several years (see chapter 6). To take better advantage of the availability of fish scrap, the company moved its works to Woods Hole, Massachusetts where it hoped to enter the fish oil business and use the scrap in its fertilizers. The location proved unsuitable for fishing operations, but the company was able to purchase all the scrap it could use from fishing companies in Maine, Rhode Island, and Long Island. In 1864 the company erected a large plant at Woods Hole to acidulate the Howland Island guano with sulfuric acid, mix it with fish scrap, and grind the mixture. (See figure 5-1.) Their promotional material claimed that the resulting product was very similar to Peruvian guano, but instead of the fish being processed and digested by cormorants and

Fig. 5-1. Woods Hole plant of the Pacific Guano Company, before 1866. (From Pacific Guano Company Collection, Case 1, Folder 7, Baker Library, Harvard University Graduate School of Business Administration.)

penguins, it was processed by machinery and digested by sulfuric acid. When the factory began operating in 1865, the product, an ammoniated superphosphate, was marketed under the brand name of "Soluble Pacific Guano." The company's trademark featured a dramatic drawing of a cormorant, with a fish dangling from its beak, standing on a bag of "Soluble Pacific Guano."[37] (See figure 5-2.)

The deposits on Howland Island were, however, nearly exhausted by 1866. The company abandoned the island in 1870, after removing 42,607 tons of guano.[38] Since the company had an extensive plant and a product with a good market acceptance, it sought alternative phosphate sources such as Caribbean phosphatic guanos. Therefore, about 1866 the company purchased the Swan Islands in the Caribbean to provide a more convenient supply.[39]

Before the company could begin to use its new guano source, however, much richer phosphate deposits were discovered near Charleston, South Carolina, in 1867. The Swan Islands became a white elephant. The company's directors attempted to find uses for the guano and may have shipped some to Great Britain, but they were ultimately forced to write off most of the cost of the two islands.[40]

Soon after the discovery of the South Carolina phosphate deposits, the Pacific Guano Company purchased mining land at Chisolm's Island at the head of St. Helena Sound, about eight miles north of Beaufort, commenced mining operations, and began shipping rock phosphate to its Woods Hole plant. Later it built a plant in Charleston, South Carolina. The company's ships carried fish scrap from Woods Hole for the South Carolina plant on the outbound trips and rock phosphate for the Woods

Hole plant on the return trip. It still marketed the product from both locations as "Soluble Pacific Guano" even though the fertilizer no longer contained any guano from the Pacific. It was, however, the custom of the company for some years to add small amounts of Caribbean guano to the product, apparently mainly for the sake of appearance.[41] The fertilizer continued to be a favorite among farmers, especially in the South, until the company inexplicably went bankrupt in 1889. The history of the company illustrates the process by which the importers of phosphatic guanos were pushed by the chemistry of their product and the economics of their situation to enter the chemical fertilizer business, which developed rapidly in the 1870s and 1880s with South Carolina rock phosphates as one of the main raw materials.

Legacy of Phosphatic Guanos

The phosphatic guanos of the Caribbean may have played an indirect role in the 1867 discovery of the rock phosphate deposits of South Carolina. The chemists who analyzed the Caribbean deposits often called them "mineral phosphates."[42] This undoubtedly facilitated the recognition of other types of mineral phosphates. The hard rock nodules of the Colombian guano, moreover, were similar in general appearance to the phosphatic nodules discovered in the Charleston vicinity. Charles U.

Fig. 5-2. Trademark of the Pacific Guano Company: cormorant standing on bag of "Soluble Pacific Guano." The company continued to use this trademark long after its products ceased to contain guano. (From Pacific Guano Company Collection, Case 1, "Forms, Labels, etc.," Baker Library, Harvard University Graduate School of Business Administration.)

Shepard, the Yale-educated chemist resident in Charleston who had been an active participant in the debate over the Sombrero guano's origin and classification, was one of the first to recognize the true nature of the Charleston deposits.[43] Moreover, one of the men credited with the beginning of production, St. Julien Ravenel, was about to commence the manufacture of fertilizers in Charleston using Navassa guano as the raw material, when he was shown some fossil and rock specimens taken from nearby Goose Creek. He immediately recognized the similarity in the material and substituted the local phosphatic nodules for the Navassa guano.[44] Thus, the similarity of the rock phosphates to Caribbean guano may have played a role in opening the South Carolina deposits that served as the mainstay of the United States fertilizer industry until the opening of similar deposits in Florida and Tennessee later in the century.

The success of phosphatic guanos as an ingredient in superphosphates may have encouraged the acceptance of superphosphates made from rock phosphates. The success of "mineral" phosphates helped dispel the idea held by farmers that much of the value of "dissolved bone" came from its gelatin and other nitrogenous material. Therefore, when the South Carolina rock phosphates were substituted for bone and phosphatic guanos by most manufacturers of superphosphates, the change did not upset most farmers.

By the 1870s the importance of phosphatic guanos had diminished. Importations continued for the rest of the century, peaking in the 1880s, but meanwhile the amount of South Carolina phosphate mined grew rapidly. By 1869 South Carolina production passed the tonnage imported from the guano islands. By 1880 South Carolina produced over two hundred thousand tons, about ten times the amount of phosphatic guano imported.[45]

Nevertheless, phosphatic guano provided some of the essential groundwork for the agricultural chemical industry that developed rapidly in the 1870s and 1880s. It conditioned agriculturalists to accept nonorganic, odorless fertilizers; it helped support some aspects of Liebig's mineral theory; it encouraged entrepreneurs to manipulate and manufacture fertilizers; and it may have played a role in the discovery of the South Carolina deposits of rock phosphate. Many of the early fertilizer companies began in the phosphatic guano trade and the fertilizer industry that developed in Baltimore and elsewhere owed part of its origin to these scattered guano islands. Although most of the participants in these developments were only seeking similar substitutes for an already existing product and there was little conscious innovation, the shape of the fertilizer industry was altered significantly by the guano island episode.

CHAPTER VI

Animal and Fish Guano

"We traverse sea and land, and send to Africa and to South America, to bring home elements of fertility which at home we throw away."[1] So complained the editors of the *Plough, Loom, and Anvil* in 1850. Indeed, entrepreneurs were importing large quantities of Peruvian and phosphatic guano, while at the same time many potentially valuable urban animal and fish refuse materials were not being used as fertilizer. The urban-rural waste recycling system that had been in place for several decades along many parts of the East Coast handled manure, street sweepings, bone, and some other kinds of solid waste relatively efficiently. But potentially more valuable materials such as slaughterhouse refuse and fish scrap were still not recycled as fertilizer. Their bulk and poor keeping qualities made them difficult to transport long distances and use efficiently as fertilizers, even though agricultural experts had long praised their fertilizing qualities.

One of the primary reasons for the emergence of guano was that the original urban-rural recycling system was not capable of supplying farmers with all of the fertilizers they needed. The Peruvian, phosphatic, and manipulated guanos had demonstrated the efficacy of highly concentrated fertilizers, dramatically changing the way farmers thought about fertilizers. Serious problems with the supply, price, and reliability of guanos, however, led to a search for substitutes. Ironically this search led directly back to the urban-rural recycling system for ways to process fleshy wastes into concentrated fertilizers similar to guano.

Animal Fertilizers

Prior to the mid-1850s, farmers seldom used blood and other animal refuse as fertilizers, although large quantities were available in all urban centers. John P. Norton, Yale's first agricultural chemist, had discussed

83

blood and flesh in a series of articles on "neglected manures" for the *Cultivator* in 1850.[2] The principal problem, as Norton recognized, was that animal waste tended to putrefy rapidly. Thus, collection and use were difficult. He could only suggest that the material should be composted. Joshua Horner, a Baltimore bone dealer, offered a fertilizer based on this principle in 1851. He added plaster, charcoal, and small amounts of lime, soda, and ashes to boiled animal flesh and then composted the mixture for six to nine months before offering his "Prepared Animal Manure" as a substitute for Peruvian guano or bones.[3] The composts, however, were bulky, low-analysis products containing large amounts of water and relatively worthless additives. They were not satisfactory substitutes for Peruvian guano.

In the early 1850s, the potentially valuable nitrogenous waste products from slaughtering were generally disposed of in the most convenient way. The operators of the largest slaughterhouses in the major cities sold some of their blood to sugar refiners and other manufacturers who used it for industrial processes, but most slaughterhouses allowed their blood to run into the gutters and neglected fleshy refuse entirely.[4] Norton described one case where the owners of a slaughterhouse allowed the offal to accumulate "year after year" in a hollow into which it was thrown.[5] Apparently the difficulty of collecting waste from the numerous small slaughterhouses and the lack of a feasible method to perserve the nutrients in a concentrated form prevented the use of animal wastes by farmers.

In the mid-1850s, after the agricultural journals began describing the British manufacture of "nitro-phosphates" from dried blood and offal, a few American entrepreneurs attempted to produce concentrated animal fertilizers.[6] The first such attempts were based explicitly on British experience. In 1856 Thomas D. Rotch, an Englishman of American descent, claimed to be the principal holder of American patents to manufacture fertilizers, according to the British method, from blood, offal, sulfuric acid, and phosphate of lime.[7] He proposed to set up plants in Boston, Philadelphia, New York, and Baltimore to use the waste from those cities. Little came of Rotch's efforts, although the *American Farmer* ran an advertisement in 1857 for a manure made according to his specifications, and a group of Boston entrepreneurs spent $40,000 to $50,000 on a plant to process butchers' offal with sulfuric acid to produce "cheval guano," so named because it supposedly was partly made from dead horses.[8] Farmers found this fertilizer useless and the operations soon failed. In 1858, however, advertisements for a similar produce called "Rothwell's English Patent Blood Manure" claimed that it had been used in England for three years.[9] Agricultural warehouses in New York advertised blood and wool manures for several years, but their products never gained wide acceptance.[10]

By the end of the decade several other companies had entered the animal fertilizer business. In 1857 John A. Schwager of New York advertised that he had arranged to obtain all the dead animals, blood, and offal of New York City. From these he produced "Empire Animal Fertilizer" at a factory on Barren Island, at the entrance to Jamaica Bay.[11] Barren Island was an ideal location for such businesses because it was within easy transportation distance from the city, but was sufficiently remote from populated areas that offensive odors did not cause problems. A rendering plant had been operating on the island since 1845 and at least one factory may have produced fertilizer from animal wastes there in the 1850s. In subsequent decades the island became a major center for fertilizer manufacture.[12] Other entrepreneurs in greater New York also entered the animal fertilizer business. The same year that Schwager advertised his "Empire Animal Fertilizer," the Brooklyn Fertilizer and Manufacturing Company offered an "Ammoniated Tafeu" prepared from night soil, blood, and butchers' offal.[13] A year or two later a New Jersey company offered a similar fertilizer composed of "dead animals, blood, offal, and raw bones.[14]

Interest in this type of manure continued in the 1860s, especially around New York, where the supply of waste was greatest. The most prominent new fertilizer introduced was "Bruce's Patent Concentrated Manure." According to advertisements taken out by two of the leading New York agricultural warehouses, this fertilizer contained "animal fibre," blood, ground bone, and an absorbent which accounted for one-fifth of its weight.[15] The Excelsior Poudrette and Fish Guano Works introduced another similar "guano" in 1866.[16]

Initially farmers were interested in these animal fertilizers for two reasons. First, these organic wastes fitted well with the prevailing mentality of recycling, since, like stable manure and ground bones, they were produced from agricultural products that farmers had shipped into the cities. Returning these wastes to farmers' fields seem a natural way to prevent soil exhaustion. Animal fertilizers also attracted attention because they appeared to be low-cost substitutes for Peruvian guano. Both chemical analysis and common sense indicated that animal fertilizers contained the same nutrients as Peruvian guano, although the price was often only one-fourth to one-half as much.

These animal fertilizers apparently did not live up to expectations. If produced in the manner claimed, the concentration of nutrients in the early products would have been low because of the large amounts of relatively worthless absorbents introduced to put the product into a satisfactory physical condition. Thus, farmers probably found these fertilizers ineffective and abandoned them after a few trials.

Even though many of the initial companies failed, interest in animal and vegetable fertilizers did not slacken. In the following decades new concerns entered the field and perfected methods of processing these materials. One factor which forced these developments was the pressure of public health authorities to eliminate the nuisances caused by slaughterhouses in the cities. In 1866 there were approximately 250 slaughterhouses in New York and Brooklyn, slaughtering over a million-and-a-half animals per year.[17] Since many of these establishments were located in populated neighborhoods and since few disposed carefully of blood, offal, or other waste, public health authorities considered them major nuisances and health hazards. Similar problems existed in other cities. In 1869 health authorities in New York prohibited slaughterhouses south of 40th Street. This led to the concentration of most of the butchers into three large abattoirs. A similar Massachusetts law passed in 1870 forced all the butchers of the Boston area into the Brighton Abattoir.[18]

The scale of operations of these new abattoirs made the collection and processing of wastes much more feasible. Fertilizer establishments soon sprang up in conjunction with many of the abattoirs. One of the first, the Manhattan Manufacturing and Fertilizing Company, located its plant next to an abattoir in Commanipaw, New Jersey. In 1871 this company began producing its "Phosphatic Blood Guano" from the blood and bone wastes of the abattoir. Its lurid trademark, showing a skull-and-crossbones on top of a heart dripping blood on the fertile field below, dramatized the recycling relationship between its animal fertilizers and agriculture. (See figure 8-6.) Its name suggested that this fertilizer was a substitute for pure guano. The Butcher's Slaughtering and Melting Association of the Brighton Abattoir manufactured a similar product which it claimed was "richer than guano."[19] Another establishment in New York first treated bones with high-pressure steam to remove fatty substances and then ground those bones not suitable for making "ivory."[20] The New York and Brighton manufacturers mixed the ground bones with finely ground dried blood and refuse meat and added small amounts of potash to produce a complete fertilizer.

Gradually, utilization of animal waste became an important part of the fertilizer supply system. Manufacturers perfected methods to dry blood and process other wastes. In the 1870s firms such as Swift & White with plants on Barren Island and in New Jersey produced complete lines of fertilizers with dried blood, meat scrap, and other animal refuse as principal ingredients, and the Kearney Chemical Works offered its "azotin," a fertilizer made from dried meat.[21] A decade later Williams & Clark in New York, which was closely associated with a large abattoir and tallow-rendering establishment, could process one hundred tons of slaughterhouse waste per day at its Staten Island plant.[22] The prominent role of slaughterhouse waste in the fertilizer industry continued to grow for the rest of the century, ultimately becoming so important that the great meat-packing firms, Armour and Swift, established fertilizer subsidiaries which were among the nation's five largest fertilizer producers.[23]

Agricultural experts welcomed these new fertilizers, primarily because animal fertilizers were an example of the kind of recycling of organic wastes they had been recommending for many years. The Brighton Abattoir advertised that its fertilizer was "nothing new," and invited farmers to tour its facilities to see for themselves that nothing except animal wastes and potash salts was used in the fertilizers. Well-made animal guanos undoubtedly were effective fertilizers.

Much of the blood and animal fertilizers, however, were not used directly by farmers. Instead, manufacturers of superphosphates, who were setting up operations about the same time, began using these waste fertilizers instead of Peruvian guano to ammoniate their products and thus produce a more complete fertilizer. Probably the first to do this was Charles De Burg, an early Brooklyn fertilizer manufacturer who began using animal wastes in his superphosphate about 1857.[24]

This gave his fertilizer an usually high percentage of nitrogen and may have contributed to its success. The stench emanating from his Williamsburg factory, however, was so offensive that city authorities eventually compelled him to stop using animal wastes.[25] Nevertheless, other companies soon followed De Burg's lead and introduced dried blood and offal into their complete fertilizers. By the 1870s many of the leading fertilizer producers in the Northern part of the country were using at least some animal wastes in this way.

Fish Guano

The fish guano industry developed from another effort to find a domestic substitute for Peruvian guano, although the roots of the industry went back to half a century earlier. Since the late eighteenth century, farmers along the shores of Long Island Sound and Narragansett Bay had used large quantities of menhaden fish on their fields. The fish were effective fertilizers, but four or five tons were needed per acre.[26] Because of the quantity needed and the rapidity with which fresh fish putrefied, it was not feasible to cart them more than a few miles from the shore. Even those farmers near the shore found that they spent excessive amounts of time in the late spring and early summer catching and carting fish.

As early as 1844 the Boston chemist Charles T. Jackson suggested that fish would make a good substitute for guano.[27] He reasoned that since Peruvian guano was basically fish which had passed through the intestines of birds, farmers using fish directly would be getting the same fertilizing ingredients.[28] A few years later, English agricultural experts also concluded that fish would make a good substitute for imported guano.[29] The major problem, as Yale's agricultural chemist, J. P. Norton, pointed out in 1849, was to find an economical way to transport fish more than five or ten miles from the coast.[30] Jackson, following the inclination of most agricultural experts for preserving manures subject to rapid decay, suggested composting the fish. This would prevent the loss of nutrients, but it increased the bulk of the fertilizer.[31] Although guano had taught agriculturalists the utility of concentrated fertilizers, it was not until the inception of the fish oil industry in the 1850s and 1860s that it was possible to produce a concentrated fish fertilizer commercially.[32]

Stories later circulated that about 1850 Mrs. John Bartlett, a fisherman's wife from Blue Hill, Maine, discovered menhaden oil while boiling some of the fish for her chickens. Supposedly, she skimmed the oil which had risen to the top of her pot, bottled it, and showed it to a Boston merchant who promised her eleven dollars a barrel for as much as she could produce.[33] Other accounts traced the beginnings of the menhaden oil industry to the War of 1812, when whale oil was scarce. To produce a substitute, fishermen processed the menhaden by allowing them to rot in barrels around Narragansett Bay until the oil rose to the surface.[34]

Whatever the origins of the fish oil business, no one combined the production of oil with the manufacture of a concentrated fish fertilizer until the 1850s, when growing scarcity and consequent high prices of whale oil encouraged entrepreneurs to seek substitutes.[35] This search led not only to Col. Edwin L. Drake's pioneering oil well, but also to experiments in extracting oil from fish.[36] In 1849 a Mr. Lewis conducted some experiments near New Haven with fish oil and manure production. His experiments attracted attention from Norton at Yale and from the British agricultural expert, James F. W. Johnston, who toured the United States in 1850.[37] By 1850 fishermen on eastern Long Island and along the Connecticut shore of Long Island Sound were extracting oil from fish by primitive fermentation methods, but there is no surviving evidence that the waste from these small-scale operations was used as a fertilizer.[38]

D. D. Wells of Greenport, Long Island, established the first commercially significant fish oil and fertilizer operation in 1850. Initially Wells partially cooked his fish and then allowed the mixture of fish, oil, and water to stand for weeks while the fish decomposed and more oil came to the top. The oil produced was very dark and foul. The horrible stench resulting from the process caused neighbors to obtain an injunction against him for "corrupting the public health." Wells then moved his operation to a more remote location on nearby Shelter Island where he set up a more efficient factory in which he first steamed the fish in large pots, and then squeezed out the remaining oil in a screw press. He sold the scrap, or "chum," to local farmers for fertilizer.[39]

Meanwhile, entrepreneurs on the Connecticut side of Long Island Sound were also entering the fish guano business. A New Haven company offered for sale in 1851 a "fish guano" that John Norton at Yale compared favorably with Peruvian guano.[40] A year or two later William D. Hall conducted some experiments with steaming and pressing menhaden at his bone-boiling and tallow-rendering establishment in nearby Wallingford. Hall patented his process for steaming fish, although D. D. Wells later claimed priority, and then Hall founded the Quinnipiac Fertilizer Company, which eventually became one of the leading producers of fish guano.[41]

The agricultural press showed little interest in these struggling enterprises until 1854 and 1855 when several journals carried accounts of pioneering efforts in France and Great Britain to produce a commercial guano substitute from fish.[42] According to these accounts, in 1851 or 1852 a Frenchman named de Molon set up a manufactory at Concarneau on the west coast of France to manufacture "guano" from the refuse of the sardine industry.[43] He and his associate Alexander Thurneyssen built a similar factory in Newfoundland to process the refuse of the cod fishery into fertilizer. The central step in de Molon's process was the steaming of the fish refuse in a revolving chamber under pressure to assure thorough cooking. He then pressed the cooked fish to extract the oil.[44] By 1856 plants using de Molon's process had been set up in Norway and England.

Two Englishmen, Pettit and Green, developed and patented a second process for converting fish into oil and fertilizer.[45] They employed sulfuric acid to break down the tissue of the fish so that the oil could be extracted. The idea of acidulating fish may have appeared similar to the treatment of bones with sulfuric acid to produce superphosphates. The fish tissue, however, resisted the action of sulfuric acid and the process was not a success, although subsequently several other entrepreneurs attempted to employ it as an alternative to cooking fish.[46]

Accounts of de Molon and of Pettit and Green, and other European efforts to produce concentrated fish fertilizers not only aroused interest among agricultural editors, but also led to a few more attempts by entrepreneurs to enter the business.[47] In 1855 S. B. Halliday organized the Narrangansett Manufacturing Company in Providence, Rhode Island.[48] The company produced a fish compost by cooking the fish, pressing out the oil, and then partially drying the residue, probably in the open air. During the process the manufacturers added gypsum, limestone, and sometimes sulfuric acid.[49] They then mixed the treated fish residue with an absorbent "in itself a valuable fertilizer," apparently street sweepings.[50] When S. W. Johnson analyzed the product at Yale, he found it contained 53 percent sand and other insoluble matter and 28 percent water.[51] Johnson concluded that despite the manufacturer's claims that this was "a very efficient fertilizer" of "great strength," it actually was worth no more than horse manure.[52] The company also produced a more concentrated fish guano, "similar to Peruvian guano," but selling for only $45 per ton.[53]

The manufacturers cooked and prepared the fish as for the compost, but instead of adding an absorbent, they dried the residue in an oven and ground it to a fine powder. Even Johnson agreed that this was a valuable fertilizer.[54]

In a related development, in 1855 a New Jersey company attempted to manufacture a concentrated manure called "cancerine" from horseshoe crabs.[55] This enterprise was modeled after a similar one in Norway which used a small sea crab as its raw material.[56] Professor Johnson did not analyze the product, but he pointed out that the analysis provided by the manufacturers could not possibly be accurate since it indicated more ammonia than was possible in animal matter. The New Jersey company apparently did not survive, although for several decades other companies attempted to use horseshoe crabs to make fertilizers.

In late 1857 a group of Long Island and New York entrepreneurs founded the "Long Island Fish Guano and Oil Works" at Southold.[57] This company advertised that it was the owner of American rights to the patents taken out by de Molon and Thurneyssen in France.[58] The entrepreneurs equipped their Southold factory with de Molon's rotating steam cooker, a hydraulic press, pickers, a drying room, and grinders.[59] According to the company's promotional pamphlet, their process produced a dry powder which contained all the constituents of the best Peruvian guano. S. W. Johnson and the editors of the *American Agriculturist* examined the product and hailed the process as the first feasible method for turning America's vast resource of fish into a concentrated manure.[60]

Soon entrepreneurs built additional factories on eastern Long Island, along the Connecticut shore of Long Island Sound, and around Narragansett Bay.[61] The development of the industry in these areas was directly related to previous use of fresh fish as fertilizers. Some of the farmers who had fished to supply their own fields and fishermen who had sold fish directly to farmers began selling fish to the new "pot works" and fish factories.[62] Farmers in the area were accustomed to using fish on their fields, so there was a ready market for the scrap. Many of the entrepreneurs came from other commercial and nautical enterprises, but some were farmers like William Glover of Southold, Long Island, who in the 1870s managed a small factory while he continued to farm.[63]

The high price of oil during the Civil War led to the rapid expansion of the industry.[64] Entrepreneurs built many new factories in the established areas of eastern Long Island, Connecticut, Rhode Island, and Barren Island, which by the 1870s was a center of fertilizer manufacturing with fish factories, rendering plants, and superphosphate manufacturers.[65]

By the end of the 1860s the industry was well established along most of the East Coast north of North Carolina. A few small-scale manufacturing attempts had been made in Maine in the beginning of the 1860s, but local entrepreneurs did not enter the business in significant numbers until the owners of Rhode Island and Long Island factories extended their operations there in 1864.[66] Likewise, a Long Island operator inaugurated the business in Chesapeake Bay.[67] By 1867 there were about one hundred factories along the eastern seaboard from New Bern, North Carolina, to Mt. Pleasant Bay, Maine.[68] The industry annually produced 30,000 barrels of oil worth about a half million dollars and about 20,000 tons of guano worth approximately the same amount.[69] By 1869 seventeen factories on the shores of Peconic and Gardiners bays on eastern Long Island produced over 7,000 tons of fish guano from more than 71 million fish.[70]

As the industry developed, the size and complexity of factories grew.[71] In the larger factories automated tramways carried fish from dockside to steam-heated cooking vats on the second floor which could handle 20,000 fish at a time. Hydraulic presses capable of a pressure of seventy tons squeezed out the oil and water. Gradually flame drying replaced the outdoor cement or wooden scrap drying platforms the early factories used. A few firms built floating factories on old ferries or barges which could be moved to follow the catch or escape the wrath of local officials.[72] By 1870 a fair-sized land-based factory cost about $70,000, although small factories could still be built for as little as $2,000.[73] While the equipment was larger and more elaborate, the basic process of cooking the fish, pressing out the oil, and drying the scrap remained the same. Since the fishing was seasonal and the oil yield was best in the fall, the factories operated only part of the year. Because of the part-time, marginal nature of the operations and because of the uncertainties of the catch, entrepreneurs were reluctant to invest capital in substantial plants and equipment. Most of the factories were cheaply constructed of wood and were often victims of fire and storm. (See figure 6-1.)

Factories were forced to locate as far as possible from populated areas because of the "vile smell" they produced.[74] Barren Island and the remote sand dunes surrounding Napeague Harbor on Long Island were typical sites for clusters of plants. Floating factories offered their owners the opportunity to escape whenever local authorities became too disturbed. For instance, in 1869 a floating factory from Old Lyme, Connecticut, operated in Oyster Bay Harbor, leaving local residents "holding their noses" until town supervisors and justices ordered it removed. Whenever a factory attempted to locate too near a populated area, town or city officials quickly took action to have the "intolerable nuisance" removed.[75] In 1871 Norwalk authorities instigated a court case against Enoch Coe's fish guano operation in Norwalk Harbor. In 1872 the Shelter Island Camp Meeting Association attempted to drive the fish guano factories from their island. Even the Pacific Guano Company felt pressure from its neighbors.[76] The entrepreneurs who eventually located their plants on Napeague Harbor called the site "Promised Land" because they had been hounded from place to place for so long.[77]

Fig. 6-1. Factory, fishing operations, and interior views of a Greenport, Long Island fish oil and guano company. (From *American Agriculturist* 27 [1868]: 452.)

The production of fish guano probably would not have been undertaken except for the profits from the sale of the oil.[78] What happened to the oil, however, was a "mighty mystery", as a Sag Harbor newspaper noted in a description of a local plant.[79] Little, if any, ever appeared in the market as menhaden oil. Instead it was generally used to adulterate or substitute for other oils such as linseed oil (used in outside painting) and neat's-foot oil (used by tanneries for currying leather).[80] The oil was also sometimes used in mining lamps and in producing "oil soap" for cleaning wool.[81] Some of the best quality oil was sent to Europe where it allegedly became "olive oil."[82]

The fishing operations that supplied these factories developed along with the processing. Earlier in the century, most menhaden were caught in shore-based seines. About midcentury, purse seines allowed a boat crew to encircle an entire school of fish with the net, close the bottom, and then scoop the trapped fish into the boat.[83] Gradually fishing "rigs" increased in size until they included at least one fast sailing "yacht," two "carryaways" designed to transport fish to the factory, and two seine boats. Often boats were owned by factories, but, as in most fisheries then, the captain and crew worked on shares. Building and outfitting the fishing fleet for a factory was generally more expensive than building the factory. Thus a $70,000 factory, such as the one Luther Maddocks built in 1870 in Boothbay Harbor, Maine, required an additional investment of $100,000 in boats and fishing equipment.[84]

A considerable quantity of the fish guano was sold to farmers in the immediate vicinity of the factories, especially those on Long Island, Connecticut, and Rhode Island.[85] Farmers appreciated the low price, which was sometimes only $15 per ton for raw scrap, and $23 to $30 for cured scrap. Because they could take the scrap directly from the factories, they felt reasonably assured that the product was not adulterated.[86] Fish had been used for decades by many farmers near the coast, so the use of fish scrap was not a drastic innovation. As the diary of one Long Island farmer shows, when the catch from his fishing company failed in 1847, he simply substituted the processed fish scrap.[87] Fish scrap also fitted the recycling mentality. It was an organic waste and had the strong smell which farmers had long associated with powerful fertilizers.[88]

The most important reason for the ready acceptance of fish scrap was that it appeared to be an excellent substitute for Peruvian guano. Much of the initial interest in the early 1850s had been aroused because of the scarcity and high price of Peruvian guano. Fish scrap seemed to be the perfect substitute. If it could be adequately dried, its composition was roughly equivalent to that of the best Peruvian guano.[89] It was potentially high in nitrogen and it had a respectable amount of phosphorus. It even looked like Peruvian guano. The almost universal use of the term "fish guano" to describe fish scrap further demonstrates the extent to which farmers and others considered fish scrap a substitute for guano.

The use of fish scrap fitted nicely into the expanded recycling system that included guano. Some editors, promoters, and farmers had justified the importation of guano by pointing out that such importations completed the nutrient cycle which began when crops were removed from the farm and sold for urban consumption. They reasoned that rivers washed the wastes from human consumption into the oceans, where they passed through the food chain to fish and eventually to guano-producing birds. Proponents of fish fertilizer argued that its use merely made the recycling more direct.[90] As the 1878 report of the Association of the Menhaden Oil and Guano Manufacturers of Maine noted, "What the sea fowl required ages to effect, the Menhaden fisherman, cooperating with the manufacturer, accomplishes . . . in a day."[91] The use of fish guano also eliminated the need for long-distance transportation and for "tribute" to foreign governments. Moreover, the use of fish guano seemed to be an adequate solution to the problem of how to use urban wastes which had troubled so many agricultural experts for years. As S. W. Johnson explained in an 1856 article in the *Cultivator*, instead of attempting the difficult task of saving the night soil from large towns, it was cheaper "unreservedly" to give city refuse to the sea and take fish "in exchange."[92]

As was the case with animal fertilizers, more fish guano was eventually used in mixed fertilizers than was sold directly to farmers. The Maine Manufacturers' Association estimated in 1878 that nine-tenths of the fish guano produced in that state was sold to superphosphate manufacturers who used it to provide nitrogen in their complete fertilizers.[93] Connecticut, Rhode Island, and Long Island fish guano factories sold a larger portion of their scrap to farmers for direct application, but the majority of their production also went into mixed fertilizers.[94] Fertilizer manufacturers generally mixed about three parts superphosphate with one part fish guano. By 1869 the Mapes company had its own fish guano factory on eastern Long Island.[95] They sold a "prepared fish guano," but probably used most of the fish in their "nitrogenized super-phosphate."[96]

Several companies were set up specifically to take advantage of fish scrap as a source of nitrogen. The Pacific Guano Company, the largest producer of superphosphates in the 1870s, located its main plant at Woods Hole, Massachusetts, because it was convenient to the menhaden fisheries.[97] Initially some of the owners of the company also held interests in fishing companies.[98] The company gave up its own fishing operations after about five years, but its records show that it continued to purchase large quantities of scrap from fish guano companies on Long Island and in Rhode Island.[99] In 1873 to 1874 the Cumberland Bone Company moved from near Portland to Boothbay, Maine, where it could obtain abundant supplies of fish scrap from Luther Maddock's adjacent Atlantic Oil Works.[100] The Cumberland Bone Company mixed the fish guano with superphosphates produced from bones from Boston slaughterhouses and rock phosphate from South Carolina. The Quinnipiac Fertilizer Company of New Haven followed the reverse course. It began as a fish guano company and then began producing superphosphates in 1871, utilizing the fish guano as a source of nitrogen.[101]

The production and distribution of fish guano encountered many of the same difficulties that plagued other types of fertilizers. Fraud was unlikely when farmers purchased their fish guano directly from a factory but many farmers purchased it through the same commercial channels as they purchased other fertilizers. Thus fraud was a possibility. Like adulterated guano, poor quality fish fertilizer was hard to detect, since considerable amounts of impurities could be introduced without diluting the characteristic pungent smell that farmers assumed was a badge of quality.[102] Fraud was especially likely in years with light catches such as 1867, when the editors of the *American Agriculturist* warned that farmers might find "a good deal beside fish refuse" in their fish guano.[103] A few years later, one analysis revealed 46 percent sand and 17 percent water in a sample of fish guano.[104] In addition to willful fraud, some of the guano may have inadvertently contained large quantities of water and other impurities, especially in the early days of the industry.

The major instability in the fish guano industry was caused by the fish. The habits of the fish not only were seasonal, but varied unpredictably from year to year. Some years very few fish were sighted, other years the catch was prodigious. Many factories were at best marginal operations, and the variations in the catch only made matters worse. After 1878 menhaden ceased frequenting Maine waters and the industy there died.[105] Everywhere the industry experienced considerable entrepreneurial turnover. Factories were often built and then abandoned after a few years, or closed sporadically when catches declined. These problems were compounded as overfishing, after introduction of effective large purse nets and steam rigs, drove the schools of fish farther from shore and exhausted some areas.[106]

While fish guano was a successful substitute for Peruvian guano and an effective nitrogenizing agent for superphosphates, the supply was limited. By 1877 the production had reached 55,000 tons but this was only one-third of the amount of Peruvian guano imported during the peak years of 1854 and 1855.[107] Production grew very little during the rest of the century. The menhaden fishery has continued to this day, but other uses for the fish, especially as an ingredient in cattle feed, have effectively priced fish scrap out of the fertilizer industry.[108] Thus, fish guano, like animal fertilizers and Peruvian guano, failed to provide the quantities of nitrogen needed by American agriculture, although the three together helped supply the limited demand which had developed by the 1870s and provided evolutionary links to the modern fertilizer industry that emerged in the next few decades.

The fish and animal fertilizers developed in the 1850s and 1860s were logical substitutes for guano. They were similar in composition to guano, but at the same time they were also similar to some of the urban wastes previously recycled as fertilizers, except that they were more concentrated. Some of the arguments merchants had used to justify guano as the reuse of urban wastes lost into the ocean, applied even more directly to fish guanos and meat fertilizers which presented a significant shortening of the waste reuse cycle. In essence the recycling system had stepped in to provide suitable concentrated fertilizers to replace guano exactly as the editor of the *Plough, Loom, and Anvil* recommended in the quotation that begins this chapter.

CHAPTER VII

Superphosphates

Neither natural guanos nor the artificial guanos manufactured from animal and fish wastes provided adequate supplies of concentrated fertilizers for American farmers. Guano supplies were quickly exhausted and the quality often was unreliable. Animal and fish guanos were partially successful substitutes, but they were also limited by supply problems. Moreover, like Peruvian guano, they were high in nitrogen but contained relatively low amounts of phosphorus, the fertilizing element most needed on depleted East Coast farms. To meet these needs, the manufacture of commercial superphosphate developed out of the urban-rural recycling system, as had the production of fish and meat fertilizers. Like these other products, the superphosphates were often called "guanos," indicating their role as a substitute for natural guano. Although the manufacture of superphosphate developed out of the recycling system, it soon evolved away from that system and eventually led to the replacement of recycled urban wastes as the principal domestic source of fertilizing nutrients.

Background

Like many of the other innovations in the production and use of fertilizers, the technological developments preceding the introduction of superphosphates into America began in Europe. Historians have not been able to determine who first dissolved bone phosphate of lime in sulfuric acid to produce a soluble "superphosphate" readily available to plants, but many attribute the introduction of superphosphates to Justus Liebig, the famous German agricultural chemist.[1]

Liebig, however, did not realize the importance of his own "discovery." He had recommended that after the bones were dissolved in acid, the solution should be diluted with water and sprinkled on fields before

96

plowing.[2] Liebig thought chemical reactions in the soil would produce a finely divided precipitation of the phosphates. He saw the process as a way to put bones in a finer state of division than could be produced by grinding. The suggestion was only one of many ideas in his *Organic Chemistry in its Applications to Agriculture and Physiology*. Liebig attached so little importance to the idea that he took no steps to exploit it or even to test it further. Moreover, when he made his own unsuccessful attempt to produce a patent manure commercially in 1845, it was not a superphosphate.[3]

John Bennet Lawes, the British gentleman farmer who devoted his life to agricultural experimentation, probably deserves credit for the development of superphosphates as a practical fertilizer. In 1841 Lawes began experimenting with superphosphates, which he later claimed to have discovered prior to Liebig's suggestion.[4] Unlike Liebig, Lawes realized the importance of his innovation. In 1843, with the aid of his long-time collaborator Joseph Henry Gilbert, he built a superphosphate factory at Deptford near London.[5] His advertisements introduced the term "super phosphate of lime" to British agriculturalists.[6] During the following decade approximately fourteen other entrepreneurs began manufacturing superphosphates in Great Britain.[7] The product rapidly gained a significant position in the British market, but remained virtually untried elsewhere.[8] In Great Britain, where large quantities of bones had been imported and used as fertilizers for several decades, the production of superphosphates was a logical step. In the United States and Germany, where the use of bones was not widespread, the introduction of superphosphates was considerably delayed.

Introduction of Superphosphates

Just as the "discovery" of guano did not lead to its application by American farmers until it fitted into their evolving fertilizer practices, the "invention" of superphosphate did not lead to the rapid adoption of this important fertilizer until successful experiences with other fertilizers had prepared the way. Word of the efficacy of superphosphates traveled rapidly across the Atlantic. J. P. Norton, the young American who had gone to Scotland to study under the eminent agricultural chemist James F. W. Johnston, and who later became Yale's first agricultural chemist, wrote to the *Cultivator* in 1844 describing the use of sulfuric acid and bones in Great Britain.[9] The *American Agriculturist* had published a recipe for mixing bone dust and sulfuric acid a few months earlier.[10] During the next several years, the agricultural press carried more recipes for making superphosphates and ran articles describing the theory and use of superphosphates.[11] Almost all of these articles reprinted items from British journals or quoted extensively from British sources. Not until 1850 did American farmers begin to experiment, apparently in response to a detailed description by J. P. Norton of how to make superphosphates on the farm.[12] Experiments prompted by Norton's recommendation generally failed because farmers attempted to dissolve the bones in acid without previous grinding, a procedure which the editors of the *Country Gentlemen* later admitted was "practically impossible."[13] By 1851 descriptions of superphosphates had become a staple item in the American agricultural press.[14]

The reports of British experience and the suggestions of agricultural experts would not have resulted in the commercial production of superphosphates without the prior success of Peruvian guano. Immediately after the introduction of guano, entrepreneurs and chemists began to develop "artificial guanos" as substitutes for real guano. At first these artificial guanos were merely physical mixtures of ingredients such as ground bone, charcoal, ashes, gypsum, and potash. The prepared guano offered from 1849 to 1854 by the New York firm of Kentish & Company was probably of this type.[15] Some, however, such as the mixture produced in 1851 by George H. Barr of New York, supposedly contained sulfuric acid along with a long list of other ingredients.[16] Manufacturers may have included sulfuric acid because they considered it an inherently valuable ingredient.[17] Superphosphates had received enough publicity, however, that at least some of the promoters of artificial guanos wanted to include (or be able to claim they had included) superphosphate in their concoctions.

Baltimore manufacturers who offered various "renovators" and "agricultural salts" as substitutes for Peruvian guano may have produced the first American superphosphates. "Chappell's Fertilizer or Agricultural Salts" and "Kettlewell's Renovator" came into the market in 1849.[18] Since the manufacturers of these fertilizers, Philip S. Chappell and Kettlewell & Davison, had been producing sulfuric acid since the 1830s, it would be logical to assume that their new fertilizers were superphosphates.[19] Early advertisements, however, made no reference to superphosphates. Moreover, when the two companies first listed superphosphates (which they called bi-phosphates) in their advertisements during the next few years, these materials were only listed as minor ingredients or novelty items that a few farmers might want to try separately.[20] Not until 1853, when Chappell introduced his "Improved Fertilizer," did either manufacturer claim to use significant amounts of bones dissolved in sulfuric acid.[21]

These Baltimore firms initially encountered stiff resistance from agricultural experts and farmers because their fertilizers were either poorly designed, carelessly manufactured, or deliberately adulterated. An early analysis of Chappell's fertilizer by the Maryland state chemist showed that it contained 27 percent sand, and the first trials reported in the press were failures.[22] Kettlewell's "Renovator" apparently encountered similar difficulties.[23] Gradually more favorable reports began to appear in the press, but apparently neither compound was successful enough to warrant continued production.[24] By the mid-1850s, Kettlewell had switched to manipulating guanos and Chappell and Davison had temporarily ceased producing fertilizers.[25]

The successful introduction of superphosphates dated from 1852 when two New York area manufacturers entered the business. The best documented of these producers was "Professor" James J. Mapes, the flamboyant and controversial editor of the *Working Farmer*, which he had founded in 1849. Mapes, a self-trained "consulting chemist," had moved to Newark, New Jersey, in 1847, where he purchased a worn-out farm and turned it into a widely admired model of agricultural improvement. In addition to his editorial duties, he performed soil analyses and advised farmers for a fee, spoke widely for agricultural improvement, and was a prominent member of the American Institute of New York City, where he was a professor of chemistry and natural philosophy. Many farmers and agricultural experts distrusted Mapes because of his "five dollar" soil analyses and his support of various dubious causes, but overall his intentions seem to have been honorable and he devoted much of his life to agricultural improvements. Although his theories were occasionally mistaken and he sometimes followed slightly dubious business practices, his teaching had a major impact on contemporaries such as William Cullen Bryant and Horace Greeley, who were Mapes's close friends. After Mapes's death in 1866, Greeley wrote that "American agriculture owes as much to him as any man who lives or has lived."[26]

Mapes was even more concerned than most of his fellow editors with the problem of securing adequate fertilizer for American farms. His *Working Farmer* devoted a considerable portion of each issue to lengthy articles on the subject, made up mainly of quotations from leading European scientists and agricultural experts such as Culhbert W. Johnson, James F. W. Johnston, Justus Liebig, Sir Humphrey Davy, or August Voelcker.[27] The journal also carried a long series of articles by Mapes on manures. Occasionally Mapes staunchly held esoteric theories, such as his belief in the "progression of primaries," whereby a mineral became more valuable as a plant nutrient the more times it was passed through living organisms.[28] By and large, however, in the early 1850s he was publishing the best agricultural knowledge available.

Because the *Working Farmer* represented Mapes's thinking so closely, it is possible to trace his venture into fertilizer manufacturing back to the urban-rural recycling system which had developed around New York City. Mapes was thoroughly imbued with the recycling mentality. In his journal, he had been a leading advocate for the use as fertilizers of all kinds of urban waste ranging from woolen rags to city sewage.[29] Among the substances he urged farmers to use was the spent bone charcoal (bone black) that had been used by sugar refineries to clarify cane syrup. In 1849 Mapes complained that New York refiners were throwing this material into the Hudson River without even experimental use by farmers even though it was as valuable a source of phosphate as ground bones.[30] In 1851 he reminded farmers that they could purchase this material for only ten cents per bushel at the sugar houses.[31] At the same time, as a devout follower of the latest British practices, he was an enthusiastic supporter of the treatment of bones by acid to produce superphosphates.[32]

In 1851 Mapes combined his two suggestions and recommended that farmers substitute bone black for raw bones in the home manufacture of superphosphates.[33] This idea did not originate with Mapes, but he was instrumental in its execution in the United States.[34] He had been employed by refineries and was familiar with the nature of their waste products.[35] Moreover, there were several refineries located around New York City within easy transportation distance of his farm.[36] Ironically Mapes's assessment of the value of spent bone black was partly based on erroneous assumptions. He recognized the importance of the phosphates, but he thought bone black was valuable mainly because its carbon would attract ammonia and its black color would help to warm the soil.[37]

Fortunately for Mapes, bone black may have been the best possible material for the manufacture of superphosphate. As many farmers had discovered, the use of raw bones often produced a viscous mass if enough acid was added to completely dissolve the bones.[38] Since bone charcoal did not contain any of the gelatinous matter found in raw bones, the problem was avoided. Moreover, the carbon in bone black acted as a conditioner to facilitate the production of a fertilizer with good physical characteristics.[39]

Initially Mapes suggested that farmers purchase the bone black and sulfuric acid and manufacture their own superphosphate.[40] In the *Working Farmer* he described in detail the method he used to prepare superphosphate for his own farm. Mapes was a strong advocate of concentrated manures and had warned farmers not to purchase manufactured "patent manures."[41] Nevertheless, in 1852 he decided to undertake the commercial manufacture of superphosphates.[42] He may have been inspired by the suggestion of the Boston chemist Charles T. Jackson that some "enterprising manufacturing chemists" should undertake the production of superphosphates.[43] Certainly Mapes was not one to ignore an entrepreneurial opportunity, as his earlier attempts to capitalize on his knowledge of soil chemistry had shown. He may simply have learned from experience that the small-scale home manufacture of superphosphates was difficult, dangerous, and needlessly expensive. As he explained when he announced his undertaking, the risk to farmers from handling sulfuric acid, the chance of breaking the glass carboys, the damage to mixing vessels, and the difficulty in obtaining sulfuric acid and pure bone dust at reasonable prices made home manufacture impracticable.[44]

Whatever factors may have motivated Mapes to become a fertilizer manufacturer, the process shows the importance of the urban-rural recycling system in producing manufactured substitutes such as superphosphates. Mapes had begun by carting stable manure from Newark. Then he substituted concentrated fertilizers such as bone for the less concentrated stable manure. Next he substituted the chemically similar spent bone black for raw bone. At the same time he began treating raw bone with sulfuric acid to produce a superphosphate. Finally he found that the bone black worked best if dissolved in acid to produce a superphosphate. From there it was only a short step to the commercial manufacture of the product. In this process Mapes gleaned suggestions from the British agricultural press, but he made only those substitutions that followed logically from his existing practices and fitted with his recycling mentality.

Mapes announced his intention to undertake the manufacture of superphosphate in the February 1852 edition of the *Working Farmer*.[45] Shortly afterward, he began producing his fertilizer at a plant at or near his Newark farm. (See figure 7-1 for an engraving of his plant about ten years later.) The first advertisement appeared in May.[46] Mapes was never shy about using his connections or his journal to promote his product. During the rest of 1853, in addition to frequent advertisements, eleven more articles in the *Working Farmer* described his superphosphate, responded to criticism, and presented the testimony of satisfied farmers.[47] Frederick McCready, the publisher of the *Working Farmer*, served as New York sales agent, and the fertilizer was sold from an office in the building of the American Institute.[48]

Mapes called his fertilizer "Improved Super-Phosphate of Lime" to

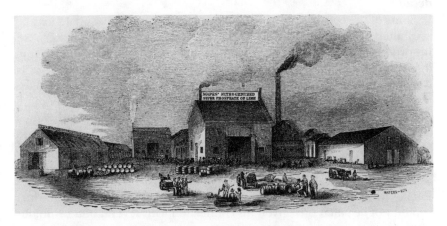

Fig. 7-1. James J. Mapes' superphosphate plant, sometime after 1859. (From *Mapes' Nitrogenized Super-Phosphate of Lime* [trade pamphlet, no date], Hagley Museum and Library, Wilmington, Del.)

distinguish it from ordinary superphosphate of lime produced in Great Britain which mainly contained phosphorus. Mapes designed his product as a complete fertilizer. The recipe he later claimed to follow called for mixing 100 pounds of bone charcoal with 56 pounds of sulfuric acid (of unspecified density), 36 pounds of Peruvian guano, and 20 pounds of sulfate of ammonia.[49] If mixed according to this formula, the "Improved Super-Phosphate" would have been an excellent fertilizer, containing up to 6 percent nitrogen and 18 percent phosphoric acid (phosphorus), although the amount of sulfuric acid used was probably only enough to transform about half the phosphates into soluble condition.[50] When Samuel W. Johnson analyzed Mapes's product in 1852, much to his surprise he found it to be an excellent fertilizer, the best superphosphate he had ever analyzed.[51] He did not trust Mapes, however, and suspected that this quality might not be maintained.[52] Later analyses by Johnson and others confirmed this suspicion.[53]

The subsequent history of Mapes's "Improved Super-Phosphate of Lime" is shrouded in controversy. Even though Johnson's 1852 analysis was highly favorable, Mapes charged that Johnson had undervalued the fertilizer and misrepresented its character. This led to an editorial duel between Mapes and Joseph Harris, the British-trained associate editor of the *Country Gentleman* and editor of the *Genesee Farmer*.[54] In a related controversy, Harris surreptitiously visited Mapes's plant and published an account charging that the mixing of ingredients was sloppy and incomplete.[55] The battles worsened during the following year when the *Country Gentleman* exposed the allegedly fraudulent manufacture of "Chilian Guano" at Mapes's Newark plant.[56] Charges that Mapes was a fraud and his fertilizer was practically worthless continued for the next decade.[57]

Some of the controversy surrounding Mapes's superphosphate may be attributed to his feisty personality. He tended to exaggerate the merits of his products and felt compelled to attack anyone who disagreed with him in even the slightest way.[58] Much of the difficulty, however, was probably due to genuine problems in the production of superphosphate. Unless the bone black had been finely ground, it was difficult to mix in enough acid to put most of the phosphates in soluble form without producing a mixture that could not be dried. He may have been compelled to add other materials as conditioners, as he had suggested before he took up commercial production.[59] Lax supervision of the manufacturing process, as Harris charged, may have made quality control even more difficult. Finally, the possibility that Mapes deliberately adulterated his product should not be ruled out, given his connection with several other dubious enterprises.[60]

Despite extensive criticism, Mapes's "Improved" must have been effective in the field, as the numerous testimonials published by Mapes claimed.[61] Worthless fertilizers did not survive in the marketplace more than a year or two. (Good fertilizers had trouble enough surviving.) Mapes, however, had to double the capacity of his plant in 1853 from ten tons to twenty tons per day to meet demand.[62] Moreover, the company survived until the 1860s, when his son Charles V. Mapes put it on a sounder footing. Therefore, the product was probably considerably better than the experts thought.

Shortly after Mapes set up his factory in Newark, Charles B. De Burg commenced manufacturing superphosphates across New York Harbor in the Williamsburg section of Brooklyn.[63] De Burg's father was a partner in a London fertilizer manufacturing business.[64] After immigrating to the United States in 1851 and setting up his own superphosphate plant in 1852, the younger De Burg claimed to have had "much experience in the manufacture of this manure in Europe."[65]

De Burg established his factory as part of the existing recycling system. Like Mapes, he chose as his raw material the waste bone black from sugar refineries which was available in the vicinity at a nominal price. Even the building he used had been part of the bone recycling system. Its previous tenant, the firm of Fryat and Campbell, had boiled bones to produce glue and then ground the bones for agricultural use.[66] De Burg apparently used the same facilities and simply mixed bone black with sulfuric acid, possibly from a nearby acid works, to produce a superphosphate.[67]

To supply nitrogen for his fertilizer, he added ammonium sulfate produced by treating the waste ammoniacal liquors of the city's gashouses with sulfuric acid.[68] Like the manufacture of superphosphate, the production of ammonium sulfate by this method had been pioneered in Great Britain and probably was part of the technological knowledge De Burg brought with him.[69] De Burg may have been the first to introduce this practice into the United States, but many other fertilizer manufacturers soon began using ammonium sulfate produced the same way.[70]

Not only did Mapes and De Burg's operations fit easily into the existing urban-rural recycling system, but they also fitted into the existing urban-centered commercial network which had developed to supply farmers with purchased fertilizers. In the major cities, agricultural warehouses such as those of A. B. Allen and Longett & Griffing in New York had been established by the early 1850s to sell agricultural equipment, seeds, and fertilizers. They provided a convenient outlet for the new superphosphate manufacturers to sell their product. Proprietors of the agricultural warehouses easily added superphosphates to their existing line of bone dust, guano, potash scrapings, poudrette, gypsum, and other commercial fertilizers.[71]

De Burg, like Mapes, often found the experts arrayed against him. An 1857 article by a "Professor Gilliam" in the *Southern Planter* concluded that as a "chemical question" neither De Burg's nor Mapes's superphosphates had "any claim to the confidence of the agricultural public," although of the two, De Burg's was "far preferable."[72] In 1852 S. W. Johnson found De Burg's superphosphate inferior to Mapes's, and in 1857 he reported that its quality had fallen off considerably.[73]

Despite these criticisms, the experts were impressed by the rapidity at which farmers had taken up De Burg's superphosphate. The editors of the *American Farmer* reported that the fertilizer was "extensively used" in Maryland.[74] Even Professor Gilliam admitted it was popular among Virginia farmers.[75] David Steward, the Maryland State Agricultural Society chemist, reported that the testimony of reliable farmers showed De Burg had produced a fertilizer which was both better and cheaper than Peruvian guano.[76] Its rapid adoption by farmers confirms that it must have been an effective fertilizer. By 1857 Maryland alone was using $80,000 worth of De Burg's superphosphates annually.[77] Apparently De Burg's factory prospered until the end of the decade, when he sold out to E. F. Coe, who continued the manufacture of high quality superphosphates at the Williamsburg site.[78]

Expansion of the Industry

Even though the first superphosphate manufacturers encountered considerable resistance from farmers and agricultural editors, other firms soon entered the field. Apparently enough farmers were satisfied with their initial experiments to create at least a small market. Moreover, entry into the business was easy. Sulfuric acid could be bought cheaply near most of the larger cities, and other raw materials were readily available. The equipment necessary to establish a factory was minimal. Grinding apparatus was useful but not essential, since some of the raw materials, such as shavings and sawdust from comb and button factories, were already in such a fine state that they needed no grinding.[79] Workmen often mixed the sulfuric acid with the bones on a flat floor or in ruts dug in the ground lined with wood or bricks.[80] The only tools needed were hoes or shovels.

Mapes later claimed that within twelve months of the introduction of his "Improved Super-Phosphate of Lime" forty other factories sprang up to produce imitations.[81] This was both an exaggeration and a bit of self-aggrandizement. In 1853 and 1854, however, four more firms in the New York City area undertook the manufacture of superphosphates.[82] (See list of companies in Appendix.) New Philadelphia firms entered the business in 1854, 1855, and 1856, and two New York firms began in 1856.[83] Most of these firms were short-lived, but one, Baugh & Sons in Philadelphia, survived well into the twentieth century. At least twenty-two more firms entered the business before the outbreak of the Civil War.[84] These were mainly in the Northeast: three in Connecticut, three in

Boston, four in greater New York, one in Providence, and four in Philadelphia. In addition, four companies produced superphosphates in Baltimore in the late 1850s and additional companies manufactured it in Charleston, South Carolina, and Richmond and Alexandria, Virginia before war broke out. Virtually all the plants were in or near urban areas. Approximately twenty-six of the thirty-six superphosphate plants established before the Civil War were still producing fertilizers in 1861.

The first entrepreneurs in the superphosphates industry were a diverse lot. Although the backgrounds of some of the early manufacturers cannot be traced, none appear to have come from agriculture. Mapes was farming at the time he established his factory, but his operation was more a demonstration farm than a livelihood. He had not been raised a farmer, but had been "bred to business."[85] Several, like Philip S. Chappell, John Kettlewell, and William Davison, entered from the chemical business. Others, such as B. M. Rhodes of Baltimore, entered from a mercantile business which had been gradually specialized in fertilizer sales and then finally undertook manufacturing.[86] At least one firm, Baugh & Sons in Philadelphia, grew out of the tannery business when the scarcity of tanbark in southeastern Pennsylvania forced them out of that business.[87]

Still other firms, such as Lister Brothers in Newark, grew out of the numerous bone grinding establishments which had begun operations in the 1840s and 1850s. Joseph Lister, who emigrated from Scotland in 1842 with his four sons, used his experience there to establish a bone button and fertilizer business in New York City that two of his sons took over on his death.[88] The company moved to Dobbs Ferry and then Tarrytown, New York. Initially the bone dust was probably only an unimportant waste product from the manufacture of bone buttons. Eventually, however, the company sold significant amounts of ground bone and then, about 1860, undertook the manufacture of superphosphate as a supplement to their main line of business.[89] Shortly afterwards the company established in Newark, New Jersey, new facilities that by 1874 had become one of the largest fertilizer plants in the country.[90] Joshua Horner, who had been grinding bones in Baltimore since at least 1850, followed the same course.[91] By 1859 he was offering superphosphates in addition to his line of ground bones and animal fertilizers.[92]

In the years immediately following the Civil War, at least fourteen more companies began manufacturing superphosphates.[93] (See Appendix for list.) Of these, five were in Baltimore, two in Ohio, and one each in Massachusetts, Maine, Rhode Island, Pennsylvania, Virginia, and Kent County, Maryland. The plants in Ohio, Maine, and Rhode Island were designed to use animal bones as raw materials. All the plants in Baltimore and the South built after the war, and most of the ones built immediately before the war, were set up to process phosphatic guano from the Caribbean.[94] Because early experiments showed that it was not an entirely satisfactory fertilizer by itself, farmers and larger entrepreneurs began mixing it with Peruvian guano to produce "manipulated guanos." The nutrients in these "manipulated guanos" were better balanced than in either Peruvian or phosphatic guanos, but the phosphates were still not in a form readily available to plants. Therefore, several of the manipulators began treating the phosphatic guanos with sulfuric acid to produce superphosphates.

Some of the new manufacturing operations grew directly out of mercantile firms with interests in phosphatic guano islands. George W. Grafflin, who later became a leading figure in the Baltimore fertilizer industry, began his career in the dry goods trade; but after he became interested in Navassa Island, he helped organize the Patapsco Guano Company to produce superphosphates.[95] Likewise, the Pacific Guano Company of Woods Hole, Massachusetts, grew out of the phosphatic guano business.[96] Its parent company was a mercantile firm which had imported phosphatic guano from the Pacific, only to discover that the material needed further processing before it could be sold.

The origins of the other plants varied. Many of the Baltimore firms were begun by men such as Gustavus Ober and William Davison, who had previously been engaged in the importation, manipulation, or manufacture of fertilizers.[97] Others were begun by men with mercantile backgrounds such as Isaac Reynolds of the Chesapeake Guano Company.[98] The Cumberland Bone Company in Boothbay, Maine, was founded by a group of farmers who, frustrated by their inability to purchase good fertilizer, decided to produce it themselves.[99] Henry Bower of Philadelphia added fertilizers to the other products produced at his Philadelphia chemical works.[100] George F. Wilson, Eben Horsford's partner in the Rumford Chemical Works, started manufacturing superphosphates to use the waste and reject bone from their main line of business, the manufacture of phosphatic baking powder.[101]

Most fertilizer factories were larger and more sophisticated after the Civil War than before. Previously, some producers had done little more than manually mix sulfuric acid and phosphatic raw materials. Newer plants had elaborate grinding equipment, power mixers, and apparatus to measure accurately the amounts of sulfuric acid used.[102] The Baugh & Sons factory in Philadelphia was probably typical of the larger operations. In 1866 its main building was 360 feet by 64 feet.[103] The basement could store 10,000 tons of raw materials and finished products. A seventy-six horsepower engine powered the main crushers, which were capable of processing fifty to seventy tons of raw bone in ten hours. The engine also powered smaller mills for grinding the bone fine, manipulators for mixing the ingredients, and the apparatus which moved the material around the mill. (See figure 7-2 for an 1867 engraving of the plant.)

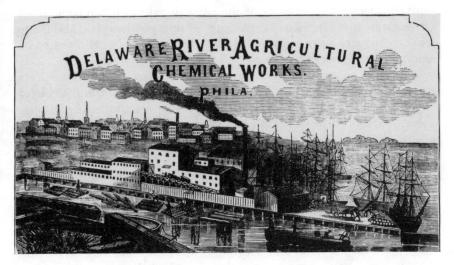

Fig. 7-2. Baugh & Sons factory in Philadelphia in 1867. (From Edwin T. Freedley, *Philadelphia and Its Manufacturers in 1867* [Philadelphia: Edward Young, 1867], 288.)

Impact of Superphosphates

Superphosphates succeeded because they could be readily substituted for other concentrated fertilizers already in use by some eastern farmers. Many farmers agreed with the claim frequently put forward in advertisements that superphosphates were ideal substitutes for Peruvian guano. A typical advertisement in the *American Farmer* for De Burg's "No. 1 Ammoniated Super Phosphate of Lime" declared in large type, "ATTENTION FARMERS! GUANO EXCELLED BY DE BURG."[104] C. B. Rogers of Philadelphia advertised that his ammoniated superphosphate was "equal if not superior" to guano on most crops and could be used in the same manner and at the same cost as guano, with longer-lasting effects.[105] Similarly, Baugh & Sons declared their "Raw Bone Super-Phosphate" was "the great substitute for Peruvian guano."[106]

Farmers making early experiments most often compared superphosphates with Peruvian guano, but also with ashes, poudrette, ground bone, stable manure, and various other substances available from the established urban-rural recycling system.[107] Clearly farmers were experimenting with superphosphates as potential substitutes for fertilizers they were familiar with, but that were too expensive or in short supply. Fortunately superphosphates could be used in the same ways as guano and other concentrated fertilizers. Thus, they were easily substituted. Moreover, the early superphosphates fitted into the recycling mentality since they were made of bone or similar material which had come from farms originally. The idea of dissolving the phosphates in acid also made sense to farmers, since most of the fertilizers they used were digested by seabirds, horses, or cattle.

The use of superphosphates spread relatively slowly at first. The press reported approximately equal numbers of successful and unsuccessful experiments (if testimonials solicited by the manufacturers are ignored).[108] Interest in superphosphates generally followed the patterns previously established by guano and other commercial fertilizers. Thus, Long Island and similar areas near northern cities were an important market, especially in the first few years. Mapes and De Burg both considered Long Island and New Jersey as prime markets when they introduced their products.[109] Baugh & Sons initially sold only to local markets in Chester, Delaware, and Montgomery counties near Philadelphia.[110]

The principal demand for superphosphates, however, developed in those areas of Maryland and Virginia near Chesapeake Bay, where the use of guano had been most firmly established. Even northern companies, which initially were set up to supply local demand, recognized the importance of the southern market. Mapes attempted to sell his fertilizers in the South.[111] Within a few years after entering the business, De Burg signed up a Baltimore agent, started to advertise extensively in the *American Farmer*, and became the leading seller of superphosphates in the South.[112] Many of the other successful superphosphate companies, such as the Pacific Guano Company and the Cumberland Bone Company, also developed extensive markets in the South.

Problems with Superphosphates

Theoretically superphosphates were an ideal solution to several problems faced by fertilizer manufacturers and consumers. Potentially they provided the best way to utilize bones, spent bone black, phosphatic guanos, and eventually rock phosphates to provide the essential phosphorus that had been badly depleted in many of the older agricultural regions of the country. Farmers, moreover, could substitute superphosphates for Peruvian guano without changing either their purchasing habits or their modes of application. In the 1850s and 1860s, however, the superphosphate industry was at best precarious. Instabilities plagued every aspect of their manufacture, sale, and use.

Even in honestly produced superphosphates, quality varied widely and was often low. Manufacturers' recipes differed considerably. Few were making pure superphosphates, which would have contained only bone or phosphatic guano mixed with sulfuric acid. Most of the manufacturers added various amounts of Peruvian guano, fish guano, slaughterhouse waste, or ammonium sulfate to provide nitrogen. Some also added a little potash. In addition, many added fillers such as gypsum, peat, or sawdust to make the product dry and manageable (or cheaper). Manufacturers, moreover, were reluctant to include the large amounts of sulfuric acid necessary to dissolve all the raw phosphates, since, under the conditions they worked in, the resulting product would have been a gooey paste. Because of insufficient acid used, a large portion of the phosphates remained insoluble.[113] The analysis of the early products was often low, compared with modern fertilizers. The five leading brands of superphosphate which S. W. Johnson examined at Yale between 1852 and 1857 averaged only 5.1 percent phosphoric acid, barely one-fourth the amount found in modern superphosphates.[114]

These unavoidable variations and deficiencies were compounded by the prevalence of fraud. In 1857 S. W. Johnson concluded that there was "no other fertilizer which so easily admits of adulteration or fraud" because "simple inspection furnishes not the slightest clue to its genuineness and excellence."[115] Appearance, taste, and smell provided no assurance of the quality of the product and often served only to mislead the farmer. The high price of concentrated manures produced what the *Country Gentleman* called a "prodigious temptation" on the part of dealers to sell adulterated products.[116] The high cost of the ingredients necessary for an honestly prepared product further tempted manufacturers to substitute inferior materials. The "quackish tone" of some of the advertisements made already suspicious farmers even more hesitant to try superphosphates. It appeared to many that "the worse the artificial manure, the louder [were] the praises."[117]

Farmers' suspicions were confirmed when Samuel W. Johnson at Yale and other agricultural chemists began publishing the results of their analyses of superphosphates. None of the values Johnson calculated in 1857 and 1858 equalled the selling prices of the product.[118] When Evan Pugh, president of the Agricultural College of Pennsylvania, carried out a similar series of analyses in 1862, the results were equally dismal.[119] Only one product had a calculated value near its selling value, but Pugh was subsequently forced to acknowledge that the sample he analyzed was not typical of the fertilizer.[120] Several fertilizers were worth only half their selling price, and according to Pugh, Mapes's was worth only one-fourth its price.[121] Given these analyses, agricultural chemists could hardly recommend commercial superphosphates, although they did continue to maintain that a well-made superphosphate would be valuable.[122]

The agricultural press was equally suspicious. The *American Agriculturist* initially was interested in superphosphates and the journal's editors eventually did recommend De Burg's and Coe's superphosphates in 1857 (while accepting advertising from the former). The editors of the *Agriculturist*, however, recommended only those superphosphates made from unburnt bones and recommended these only for their "organic matter," not their phosphates.[123] Later the editors were critical of Johnson and Pugh for granting such high values to phosphatic fertilizers.[124] In 1863 they could not advise farmers to use the product, although they continued to accept advertisements from manufacturers they considered reputable (not Mapes).[125] In 1867 the editors were still suspicious. They recognized the value of a good superphosphate, but despaired of finding one. Other journals such as the *Country Gentleman* were just as suspicious.

Farmers and their expert advisors often tried to solve the problems of unreliable quality and adulteration by avoiding commercial superphosphates. In the late 1860s experts were still recommending that farmers mix their own superphosphates, although that was a difficult and possibly dangerous operation.[126] Even more frequently, editors simply recommended that farmers not use superphosphates.[127] Experts were always ready to recommend retreat into the on-the-farm recycling system. Yale's S. W. Johnson devoted most of his attention in the late 1850s to the analysis of various mucks and peats which might provide local substitutes for superphosphates and other commercial manures.[128] Farmers often turned to alternative fertilizers which appeared to be safer, less likely to be adulterated, and consistently higher in quality. Many farmers in the South preferred manipulated guanos while many farmers in the North continued to purchase ground bones. Some farmers experimented with fertilizers made from blood, animal wastes, fish, and New Jersey marl that entrepreneurs offered as alternatives to superphosphates.

Manufacturers who attempted to produce a reliable, reasonably priced, trustworthy superphosphate faced a series of difficulties. They constantly encountered problems caused by the scarcity, poor quality, or high cost of raw materials. The supply of sugarhouse waste was limited and the material soon became more costly as demand for it developed. Raw bones had always been expensive. Manufacturers attempted to substitute phosphatic guanos, but their supply was also limited and their use created additional problems. Their quality was inconsistent, especially those from the Pacific. Some were very low in phosphates, and many contained impurities which, when acidified, absorbed considerable amounts of the sulfuric acid, produced chemicals harmful to plants, and made the product difficult to handle. Manufacturers also considered using domestic deposits of apatite and marl, but the raw material problem was not solved until the substitution of South Carolina rock phosphates in 1867. Manufacturers attempted to improve the quality of their products by using better grinding and mixing equipment and by using higher analysis raw materials, but techniques were still not satisfactory by the mid-1860s. Manufacturers attempted to shore up their reputations by using brand names, trademarks, and extensive advertising, but many farmers and experts remained suspicious. Many of these efforts to produce a reasonably priced, quality product that farmers could trust bore fruit in the following decades, but in the mid-1860s, the substitution of superphosphates for Peruvian guano and other concentrated fertilizers was still precarious.

Conclusions

In the shadow of professional criticism and practical problems, it is surprising that superphosphates were used as widely as they were. The success of some experiments was dramatic enough to convince farmers of the value of superphosphates, but probably the main reason for their limited success was that many farmers were desperate for adequate, economical sources of fertilizers. Editors and chemists could continue to extol the virtues of recycling all available local wastes, but farmers understood all too well the incredible amounts of effort involved. Moreover, farmers were anxious to seek out bargains, and most superphosphates sold at prices enough below that of Peruvian guano to make them attractive. Thus the unusual situation developed in which the agricultural experts and their scientific mentors were advising extreme caution, but practical agriculturalists were moving ahead with a risky new fertilizing material.

Theoretically superphosphates should have solved many of the problems plaguing the manufacture and use of fertilizers. Well-made superphosphates were an excellent substitute for Peruvian guano. They made the use of ground bone and phosphatic guanos more efficient and they enabled the production of properly balanced mixed fertilizers at a reasonable price, but because of the new problems superphosphates introduced into the fertilizer system, these benefits were appreciated only slowly.

Gradually, however, a viable superphosphate industry emerged from the confusion. By the late 1850s farmers reported more successful experiments, although failures still occurred.[129] The few surviving sales figures and the number of new entrants into the business show that superphosphates were gaining in acceptance.[130] Many difficulties remained, however, which made the success of superphosphates tentative until the 1870s and 1880s when the principal problems of raw materials, quality, and adulteration were solved.

CHAPTER VIII

From Chaos to Stability

At the beginning of the 1870s prescient observers such as Stephen L. Goodale, secretary of the Maine State Board of Agriculture, and Charles A. Groessman, a professor at the Massachusetts Agricultural College, could see the potential for large-scale use of commercial fertilizers in the East, but the fertilizer industry was still beset by serious instabilities.[1] Adequate raw materials were not available to supply any of the three most important fertilizer nutrients. The prevalence of adulteration and fraud in the industry was notorious. Production difficulties made the products of even honest manufacturers not entirely reliable. Moreover, farmers in only a few areas used fertilizers extensively, while elsewhere widespread suspicion or apathy reigned. Commercial fertilizers could not become a viable support system for American agriculture until all of these interconnected problems were solved.

South Carolina Phosphates

Phosphorous was the plant nutrient most needed in many long-cultivated areas of the eastern United States.[2] At the same time, its supply posed the greatest difficulties. Adventurers had explored the tropical seas in search of guano islands and entrepreneurs had ransacked the waste piles of the cities in search of recyclable phosphates; but neither had provided adequate supplies. The phosphatic guano was sometimes of high quality, but those deposits were quickly exhausted. The poorer quality of the remaining deposits and the long shipping distances from the Caribbean or Pacific limited the extent to which guano phosphates could be used. Raw ground bones and the waste bone charcoal from sugarhouses were also good sources of phosphates, but the limited quantities available led to high prices.

British scientists and fertilizer producers had used mineral phosphates

112

as substitutes for bone in the 1840s. John Bennet Lawes's original superphosphate patent of 1842 included the use of mineral phosphates. His
factory made extensive use of coprolites, the fossilized excrement of
extinct animals found in some areas of Cambridgeshire, Bedfordshire,
and Suffolk.[3] James Murray's superphosphate patent of the same year
specifically included the use of coprolites, which he was apparently
already using to produce superphosphates.[4] In 1843 the British government sent Charles Daubeny, a professor of geology and chemistry at
Oxford, to Estremadura, Spain, to investigate "phosphorite" deposits
there. Although these deposits were too far from the coast for profitable
exploitation, experiments conducted in England showed that these
mineral phosphates were as valuable as bone.[5]

Information about the English use of mineral phosphates may have
promoted the discovery of two deposits of "native phosphate of lime" in
the United States. By the early 1850s American agriculturists were familiar with the British use of coprolites as a raw material for superphosphate
manufacture. One journal, after describing the large quantities used in
Great Britain, suggested that if the American government devoted to the
search for phosphates a tiny fraction of the amount spent on weapons, it
would undoubtedly find similar valuable deposits in the United States.[6] In
1851 Dr. Ebenezer Emmons, professor of minerology and geology at
Williams College, and a supervisor for the New York Geological Survey,
reported an "inexhaustible" vein of rock phosphate or apatite near
Crown Point, New York, which he had discovered several years earlier.
He attempted to sell his phosphate rock as a substitute for guano and
bone dust, and he manufactured samples of superphosphate from his
rock, but neither attempt was successful, apparently because of the characteristics of the rock.[7]

About the same time, Dr. Charles T. Jackson, a Boston doctor who had
been involved in testing various guanos and fertilizers, and Francis Alger,
also of Boston, discovered a similar deposit in Hurdtown (near Dover),
New Jersey and suggested that some "enterprising chemist" should use
these phosphate rocks to produce superphosphates. The agricultural
press circulated his comment widely.[8] One journal suggested that these
mineral phosphates were "a cheap and practical substitute for bone and
guano."[9] Jackson and Alger distributed samples to agriculturists for free
trials and attempted to induce entrepreneurs to use these phosphate
deposits. But manufacturers found it difficult to convert the crystalline
apatites into soluble form, although the New Jersey Zinc Company may
have briefly produced superphosphates from these deposits in 1852.[10]
The promoters sent a trial shipload to England, but British fertilizer
companies found it to be of such poor quality that they left it on the dock
where it was eventually used for ballast in outgoing voyages.[11] Interest in
mineral phosphates continued, but no other deposits were discovered in

the United States for over a decade, although apatite mines were opened in Quebec and Ontario in 1863.[12]

The real breakthrough came with the discovery of the South Carolina rock phosphate deposits in 1867.[13] Beds containing phosphate nodules lay a few feet beneath the surface of considerable areas in a seventy-mile stretch of the South Carolina coast ranging from the headwaters of the Wando River north of Charleston, to the mouth of the Broad River near Port Royal, and extending inland up to thirty miles.[14] Although the discovery of the value of these deposits appeared to have been quite sudden, it grew out of developments which began forty years earlier when a few South Carolina agriculturists, like their brethren to the North, became interested in marl as a possible soil amendment. Their interest led the state of South Carolina to commission Lardner Vanuxem to make a geological and mineralogical survey of South Carolina. Vanuxem predicted in his 1826 report that marl deposits would be found there which would be as valuable as the ones then being exploited in New Jersey, but no marl deposits were discovered for fifteen years.[15]

In 1842, James Henry Hammond, the newly elected governor of South Carolina and a devoted agricultural reformer, engaged Edmund Ruffin, the Virginia farmer who was the leading proponent of marling, to do a geological and agricultural survey of South Carolina. Hammond and other like-minded South Carolinians hoped that Ruffin would find marl deposits in their state as he had done in Virginia. Ruffin did not disappoint them. Although he served for only one year, he discovered marl beds near Charleston that were supposedly three times as rich in calcium carbonate as the Virginia deposits. He enthusiastically recommended the marl to South Carolina planters, but the results from the first applications were disappointing.[16] Meanwhile, Charles U. Shepard, the well-known mineralogist and chemist, and J. Lawrence Smith, his student at the Charleston Medical College, discovered that the marl contained between 7 and 9 percent phosphates. The 1846 report of Ruffin's successor as state surveyor, Prof. Michael Toumey, provided a much more detailed description of the "calcareous strata" of the Charleston basin, including an account of Shepard and Smith's analysis.[17]

Ruffin and Toumey had both noticed in the strata above the marl beds a layer of detached nodules they assumed had broken off from the marl. These nodules were exceptionally high in phosphates, although at the time neither suspected their composition differed from that of the underlying marl layer, and Ruffin specifically rejected the nodules as useless for agriculture since they contained little calcium carbonate.[18] The numerous fossils among the nodules attracted the attention of Harvard's geologist Louis Agassiz, the British geologist Sir Charles Lyell, who had traveled through the area during the 1840s, and Francis S. Holmes, professor of zoology, geology, and paleontology at the College of Charleston. Professors Holmes and Toumey published a multivolume work, *The Fossils of South Carolina*, between 1855 and 1860. Even though they made extensive fossil collections, none of these investigators realized that the fossils and the accompanying nodules contained commercially significant concentrations of phosphates. Nevertheless, the scientists interested in fossils and agricultural marl did accumulate considerable information that facilitated the later discovery of the rich Charleston phosphate beds.

A second sequence of developments leading to the discovery of the South Carolina rock phosphates began with attempts to find substitutes for phosphatic guano in the production of commercial fertilizers. Charles U. Shepard, who had discovered the phosphatic nature of the local marl deposits, in 1856 had been among the first scientists to analyze and describe samples of the rocklike phosphatic guanos from the Caribbean when a vessel carrying one of the first shipments from Venezuela had been forced to discharge its cargo in Charleston.[19] Then in an 1859 speech before the Medical Association of South Carolina, he predicted that the supply of phosphatic guano from abroad would soon fail and that Charleston would then supply the nation with "phosphate stone" as a substitute.[20] Shepard did not specifically state that he had in mind the nodules found in the strata above the marl deposits, but he must have known that the marl itself was not concentrated enough to be commercially valuable. The "striking resemblance" between the phosphatic guano that he had examined and the South Carolina nodules, however, may have suggested to Shepard that the latter might also be a concentrated source of phosphates.[21]

Shepard's guess that the South Carolina nodules might be similar in composition to the phosphatic rock guano, which they resembled physically, led almost immediately to entrepreneurial attempts to exploit the deposits later in 1859 when Col. Lewis M. Hatch, a Charleston commission merchant, sought Shepard's assistance as a chemist in setting up a fertilizer factory in Charleston. Hatch, who was a dealer in Colombian guano, had previously engaged Shepard to examine incoming cargoes.[22] Now Hatch, along with T. P. Allen, his brother-in-law, and Melvin P. Hatch, his son, was preparing to manufacture fertilizers from bones, ammoniacal liquor, ashes, and other urban wastes from Charleston. When the operation quickly exhausted available supplies of bone, Shepard apparently suggested that they should examine the phosphatic marles along the Ashley River west of Charleston, since he suspected those deposits might prove significantly richer in phosphates than had been thought and might make excellent substitutes for bone in the fertilizer plant. The following spring, Colonel Hatch collected samples of the nodules and sent them to Shepard in New Haven. Shepard had the rock ground and applied to his garden, where the results were encouraging. Since he was away from his laboratory, however, he did not perform an analysis. Nevertheless, he was sufficiently convinced of the value of the nodules that he recommended them to Hatch in the fall of 1860.[23] About the same time, Shepard also apparently agreed to produce fertilizers from the phosphate rocks in conjunction with George T. Jackson of Augusta, Georgia, who had commenced manufacturing fertilizer there from bone waste, but had also quickly exhausted his supply and was seeking a substitute.[24]

The Civil War halted both of these budding enterprises. Jackson apparently did not attempt to revive his operation after the war.[25] Hatch moved to North Carolina, and after consulting with Shepard, attempted to sell his fertilizer plant and equipment to a partnership formed by the owner of the plantation along the Ashley River on which Hatch had found the phosphate rocks and a New Haven entrepreneur who was to supply the capital. The effort was abandoned when the New Haven entrepreneur drowned on his way south. During the Civil War, however, Hatch (so he later claimed) told his brother-in-law John R. Dukes, a partner in the Charleston factorage firm of W. C. Dukes & Sons, about Shepard's belief that the Ashley River nodule deposits were similar to Colombian guano.[26]

After the war, W. C. Dukes & Company began a new fertilizer works in conjunction with D. C. Ebaugh and St. Julien Ravenel, a Charleston chemist and physician who had earlier recommended burning the phosphatic marl for fertilizer.[27] Initially this company planned to use Navassa phosphates as their principal raw material.[28] During the summer of 1867, however, Ravenel learned of the high phosphate content of the local rock nodules, substituted these in his manufacturing plans, and forwarded to Baltimore the unused Navassa phosphates that arrived in November.

Accounts of the discovery of the high phosphate content of the nodules are confused. Ravenel later claimed that about the time he was preparing his manufactory during the summer of 1867, he happened to receive from Dr. F. M. Gedding several specimens of fossils, marl, and nodules from the latter's plantation on Goose Creek a few miles north of Charleston. Ravenel then supposedly analyzed the rocks and discovered their true constitution.[29] Ravenel's interest in the phosphatic nodules may have been aroused initially by his friend and business associate, John R. Dukes, who had learned of Shepard's opinions about the value of the rocks from Hatch during the war. Ravenel may also have learned more directly of Shepard's views. At any rate, he was interested enough to collect samples of the rock, although he did not appreciate their high quality at first. If he had originally intended to use local phosphates, his company would not have ordered the Navassa phosphate.[30]

Another participant in the discovery told a slightly different version of the story. N. A. Pratt, a Georgia-born, Harvard-educated chemist, claimed later that after the war he went to Charleston seeking sufficiently rich phosphate deposits to support a fertilizer factory.[31] His interest supposedly had been piqued by Professor Holmes, with whom he had worked during the war while in charge of the Confederate States Nitre and Mining Bureau. Holmes had been familiar with the local marl deposits since 1842, when he brought some of them to Ruffin's attention. He knew about Shepard's 1845 analysis and may have informed Pratt of Shepard's hunch that the rocks might have commercially significant amounts of phosphates. During the war Holmes had used marl found near Ashley Ferry in the manufacture of saltpeter. While digging the marl, his workmen stumbled on a pocket of nodules which Holmes thought might be coprolites. He gave some to Pratt for analysis, but the latter found only 15 percent phosphate.[32]

According to Pratt's account two years later, he happened to be visiting Ravenel's laboratory in August 1867 when the latter showed him the Goose Creek samples. Pratt claims to have discovered that the nodules contained 34 percent phosphate of lime, much more than anyone had suspected, and to have communicated this information to Ravenel. Pratt then took the sample and showed it to his friend Holmes, who immediately recognized it as one of the local marl rocks with which he had been familiar since 1839. He showed Pratt a large number of similar specimens in his fossil collection at the College of Charleston. Pratt took a few of Holmes's rocks and tested them. To his astonishment, he found they contained almost 60 percent phosphate of lime. Pratt then hurried back to Holmes, whose extensive knowledge of the local geology made it relatively easy for the two men to locate large deposits of the phosphate nodules in the Charleston area.[33]

The precise order of events in this discovery is unclear, but the impact of the various factors that led to the discovery is clearer. Investigators first accumulated knowledge about the local geology because of agricultural interest in marl. Later, curiosity about Colombian rock guano led to curiosity about similar-looking nodules found near the marl deposits. Finally, the search for possible substitutes for bone and phosphatic guano directed attention toward the local deposits. Each step of the way, scientific inquiry and the accumulation of scientific information facilitated the discovery which eventually took place, but ultimately the need to find a substitute for guano provided the incentive to experiment with the nodules.[34]

Development of the South Carolina Phosphates

Immediately after the discovery, both the Charleston parties formed companies to exploit the South Carolina deposits. Pratt and Holmes first sought local financial support and when that failed, secured the backing of "a few enterprising Philadelphians" including George T. Lewis and

Frederick Klett, who were already familiar with the chemical and fertilizer industries there.[35] On November 29, 1867, they formally organized the Charleston Mining and Manufacturing Company with Holmes as president and Pratt as chemist and superintendent. Paid-up capital was supposedly one million dollars, although this amount appears to have been far in excess of the company's actual needs.[36] Meanwhile, Ravenel and his associates, using local capital, organized the much smaller Wando Mining and Manufacturing Company with John R. Dukes as president.[37]

Both companies quickly attempted to gain control of phosphate lands along the Ashley River between five and fifteen miles northwest of Charleston. The Charleston Mining and Manufacturing Company with its larger capital acquired much more land than its smaller rival. Both. companies quickly shipped samples of their rock phosphates to northern fertilizer manufacturers for trial and further analysis. The Wando company shipped its first small sample to George E. White, a New York City fertilizer dealer, on December 4, 1867, and the Charleston company shipped samples later the same month to Potts & Klett, the Philadelphia sulfuric acid and superphosphate manufacturer with which Frederick Klett was associated. Potts & Klett apparently ran the sample through their superphosphate plant on a trial basis. The following April, a booming trade in South Carolina phosphate began when the Wando company shipped the first commercial cargo to Baltimore and the Charleston company sent two schooner-loads to Philadelphia.[38]

The Wando and Charleston companies obtained their phosphate from land mining. Laborers dug a series of small pits in the swampy phosphate beds and removed the phosphatic nodules by hand. Later, as the scale of the operations increased, the companies laid tramways through the areas to be mined and organized laborers into gangs who started long ditches between sets of tramtracks and dug in an orderly fashion toward the tracks.[39] The ditches were generally four to seven feet deep. Deposits deeper than seven feet were not mined. Each laborer dug about one ton of nodules per day, for which he received $1.75.[40] (See figures 8-1 and 8-2 for contemporary illustrations of phosphate mining.) Black laborers performed most of the work, since white laborers considered the "malarial climate" too unhealthy and the owners found the black laborers more "docile."[41] As it came from the pits, the rock was contaminated by large amounts of dirt. The companies first washed it by hand in nearby streams, but they soon saw the necessity of installing mechanized washers to produce a cleaner rock that was both cheaper to ship and easier to process.[42]

The beds of the rivers that flowed through the phosphate region were also littered with the phosphate nodules. Initially, boatmen using oyster tongs or hand-held "grabs" scavenged for rock and black divers collected the rocks from the river bottom by hand.[43] Soon, however, capitalists began to organize large-scale companies to work the "river rock" deposits.

Fig. 8-1. "Quarrying phosphates in South Carolina." (From *American Agriculturist* 31 [1872]: 20.)

Fig. 8-2. Wando phosphate mines on the Ashley River. Note washing plant in background and dock on right for shipping cleaned rock to superphosphate plant. (From *Wando Mining & Manufacturing Company* [Charleston: n. pub., 1869?].)

The first was the Marine and River Mining Company, organized in 1870. William L. Bradley, the Boston superphosphate producer, was a leading stockholder in this company, which began with a capitalization of $500,000, half paid in. The company filed the $50,000 bond required by their act of incorporation, which also stipulated royalties of $1 per ton, and began mining operations the following year under the superintendence of C. C. Coe, probably also of Boston. Meanwhile, it transferred rights (which it incorrectly believed to be exclusive) to three additional river mining companies.[44] By the end of the next decade, "river rock" accounted for almost half of the total South Carolina phosphate production.

These companies quickly replaced hand-tonging with powerful steam dredges. Some of these were "dipper" dredges similar to the ones used in ordinary navigational dredging and capable of lifting about 100 tons per day. Others were "grappler" dredges designed specifically to dislodge deposits firmly embedded in the river bottom. One company even developed a specially designed suction dredge for the work. No matter which type of dredges were used, laborers still needed to separate the phosphate nodules from the accompanying sediments by hand.[45]

Northern urban entrepreneurs played an important role in the development of the South Carolina deposits. By 1870, of the fourteen companies shipping phosphates from South Carolina, five companies, which accounted for over half of total shipments, were wholly or partly owned by fertilizer manufacturers based in northern cities. The Charleston Mining and Manufacturing Company, the largest shipper, was actually a Philadelphia corporation. William Bradley of Boston not only held an important interest in the largest river company, the Marine and River Mining Company, but he also held an interest in the Carolina Fertilizer Company, which was the second largest producer of land rock. The Pacific Guano Company, founded by Boston merchants, owned a mine on Chisholm's Island near the head of St. Helena Sound. And George S. Scott of New York was proprietor of the Oak Point Mines. Probably northern urban capitalists held interests in some of the other companies, but ownership is difficult to trace.[46]

Although at first large quantities of South Carolina rock were shipped to northern and European superphosphate factories which then shipped much of their product back to the South, the discovery of the Charleston deposits encouraged entrepreneurs to construct factories near Charleston, where they would be adjacent to the source of their principal raw material and nearer the most important markets in Georgia and the Carolinas. In November 1867 the Wando company had been the first to manufacture superphosphates in the Charleston plant shown in figure 8-3. They used acid imported from the North. The owners of the Charleston Mining and Manufacturing Company organized a separate company,

Fig. 8-3. Wando Company's Fertilizer Works on the Cooper River at the foot of Hasel Street in Charleston in 1869 This was the first fertilizer plant built in Charleston to process local rock phosphates. It had a fifty horsepower engine and a drying kiln, apparently located in the low structure at the rear of the plant. The fourth floor of the mill proper contained bins for storing raw, pulverized rock. The third floor contained the mixer and space where the finished fertilizer could dry and cool. The other floors housed ore crushers and pulverizing machines. Elevators connected the four levels. The shed on the left side was a warehouse for finished products. (Engraving from *Wando Mining & Manufacturing Company* [Charleston: n. pub., 1869?], 6.)

the Sulfuric Acid and Superphosphate Company (also called the Etiwan Company) to manufacture acid and produce "Etiwan guano" and other fertilizers. In 1868 they began construction of a fertilizer factory (shown in figure 8-4) and commenced production in 1870. It undoubtedly incorporated the latest technological advances made by the innovative Philadelphia firm of Potts & Klett. The plant also supposedly included the first acid chamber south of Baltimore, although the Pacific Guano Company may have begun producing sulfuric acid in Charleston a year earlier. By 1871 three more companies, including the Wando Company, had also constructed acid chambers to produce their own sulfuric acid.[47]

The Pacific Guano Company set up the third plant near Charleston. This Massachusetts company, which began its operations using phosphatic guano from the Pacific, had shifted to Caribbean phosphatic guanos when the Pacific deposits ran out, and then quickly substituted South Carolina rock in its "Soluble Pacific Guano." The company began

Fig. 8-4. Etiwan Works of the Sulphuric Acid and Superphosphate Company, Charleston, South Carolina, in 1870. This works included the first acid chamber south of Maryland. Note how much larger this plant, built with northern capital, was than the plant of the Wando Company, built with local capital, shown in figure 8-3. (Engraving from advertisement in Francis S. Holmes, *Phosphate Rocks of South Carolina . . . with a History of Their Discovery* [Charleston: Holmes' Book House, 1870], 2.)

mining phosphates on Chisolm's Island in 1868.[48] At first it shipped phosphates to its Woods Hole plant and shipped most of its fertilizer back to the South. Then in 1869 it constructed a plant near Charleston that could be supplied more conveniently from the Chisolm's Island mines and was also nearer the company's main markets.[49] The company continued to ship rock to its larger Woods Hole plant. Ships returning from Woods Hole sometimes carried fish scrap to ammoniate the product manufactured in Charleston.[50]

By 1873 three more superphosphate plants were operating in the Charleston region.[51] One of these, the Wappo Mills, was owned by John B. Sardy, a leading New York guano merchant who had sold Jarvis Island and Baker Island guano as an agent for William H. Webb and later had been an agent for the Pacific Guano Company.[52] Northern capitalists may also have held interests in some of the others. The influx of northern capital and know-how was undoubtedly at least partly responsible for the rapid growth of fertilizer manufacturing in South Carolina. By 1869 South Carolina was already the fifth largest fertilizer-producing state in the nation. Ten years later it became the third largest producer; by the end of the century, it had become the second largest producer, trailing only Maryland.[53]

South Carolina phosphate production quickly burgeoned from six tons in 1867 to 65,000 tons in 1879, and more than 200,000 tons before the end of the decade. In 1893, the maximum year of production, it exceeded 600,000 tons. Approximately half of the production in the 1870s was exported, mainly to Great Britain. Another 13 percent, averaging about 20,000 tons per year, was consumed locally. The other 34 percent was shipped to fertilizer manufacturers in the North. During the following decade, the consumption of phosphates by South Carolina fertilizer producers rose from 20,000 tons to about 100,000 tons per year, but the state still produced only about 15 percent of the total amount of fertilizers in the country. Almost half of the rock was then sent North for processing and the other 36 percent was shipped to Europe.[54] Nevertheless, the opening of the South Carolina deposits did initiate a significant shift southward in the fertilizer industry's center of gravity.

The South Carolina phosphates replaced phosphatic guano with amazing speed. Total production in South Carolina passed the level of importation of phosphatic guano in 1869 and by 1880 was ten times greater.[55] Fertilizer manufacturers quickly substituted the rock phosphates for guano in their superphosphates. Border state manufacturers may also have substituted the rock phosphates for phosphatic guano in their manipulated guanos.[56] The Department of Agriculture estimated in 1870 that three-quarters of the fertilizer manufacturers in the northern states were already using South Carolina phosphates.[57] The southern fertilizer industry, which developed after 1867, was based almost entirely on the acidulation of South Carolina rock. In addition, some farmers who had been using ground bone or guano directly, substituted finely ground rock phosphates. Although these phosphates were less readily available for plants than superphosphates and larger quantities were needed per acre, the cheaper price and the long-lasting effects appealed to some farmers.[58]

The South Carolina phosphate rock could not have been introduced so rapidly except as a substitute for other established fertilizer ingredients. Its successful exploitation depended on the existence of the superphosphate industry. The prior utilization of rock guanos had prepared American fertilizer manufacturers to process the hard South Carolina rock phosphates. Northern fertilizer manufacturers, whose operations had grown out of the urban-rural recycling system, played a prominent part in financing and directing the development of the South Carolina deposits and in transplanting a major portion of the industry to that area.

The recycling mentality may have facilitated the introduction of rock phosphates. Although no one attempted to equate the use of rock phosphates with recycling as they had Peruvian guano or fish scraps, early promoters of the rock phosphates were careful to stress that the deposits were entirely organic in origin. Theories about the origins of the deposits were somewhat confused, but most experts agreed that the rock phosphate deposits were some kind of animal remains.[59] Shepard even thought that the deposits might be petrified bird guano.[60] All of these theories about the origin of the rock phosphates reassured farmers accustomed to fertilizers produced from vegetable or animal wastes that these phosphates were not radically different from the organic phosphates in bones and guano which they had previously been using. The Sulfuric Acid and Superphosphate Company even called their superphosphates "dissolved bone" or "Etiwan guano" although the product contained neither bone nor guano and was made primarily of acidulated South Carolina rock phosphates.

No one intended to change the nature of the system by the substitution of South Carolina phosphates for phosphatic guano, but the effects were profound. The abundance of the deposits allowed and even necessitated the growth of large-scale manufacturing. The lower prices of superphosphates produced from South Carolina rock made the use of commercial fertilizers feasible for the larger numbers of farmers.[61] Moreover, the use of South Carolina rock altered the actual process of manufacturing and the geography of the industry. Most importantly, the substitution of South Carolina phosphates eventually ended the reliance on the urban-rural recycling system as the prime source of phosphates for American agriculture.

Other Raw Materials

The introduction of South Carolina phosphates solved the most pressing raw material need of the fertilizer industry, but other problems remained. The South Carolina discovery did not even completely satisfy the demand for phosphates. Manufacturers continued to import phosphatic guanos, especially from Navassa, for the rest of the century, possibly in part because they had a somewhat higher percentage of phosphate than the South Carolina rock phosphates.[62] Manufacturers also continued to use raw bones and spent bone black in their fertilizers, often because of lingering notions among farmers about the value of a fertilizer made from real bone. To supplement local supplies, East Coast manufacturers began to import bones from western slaughterhouses.[63] They also purchased buffalo bones collected along the new western railroads and imported bones and bone ashes from South America.[64] Nevertheless, these alternative sources of phosphates played a relatively minor role in the development of the fertilizer industry. By 1883 Baltimore manufacturers used only about 20,000 tons of bone, bone black, and phosphatic guano compared with almost 80,000 tons of South Carolina phosphates.[65]

The amount of bone and guano used was probably only slightly higher for manufacturers in the North.

Attempts to move beyond the urban-rural recycling system for nitrogenous fertilizers were generally unsuccessful in the nineteenth century. The industry continued to derive its nitrogen supplies mainly from organic sources such as processed slaughterhouse refuse and fish scraps.[66] Fertilizer manufacture developed in the Midwest in conjunction with the rapidly growing meat-packing industry there. By the end of the century fertilizer manufacturing had become an important subsidiary of the nation's largest meat-packing companies.[67] Some manufacturers continued to use small amounts of Peruvian guano, others made use of whatever ammonium sulfate was available from city gasworks, but both of these nitrogenous materials were expensive, often costing two or three times as much per ton as the phosphatic raw material.[68] A few farmers experimented with Chilean nitrates (also known as caliche or saltpeter), but until changed political conditions and the introduction of the "Shanks process" in the 1880s, these were too expensive for use in fertilizers.[69] The industry did not discover satisfactory substitutes for scarce and expensive organic nitrogen until the end of the century. Meanwhile, the fertilizer industry intensively developed the waste recycling system to provide sufficient supplies of nitrogen.

Fertilizer manufacturers fared somewhat better in their attempt to find adequate supplies of potassium, the third major fertilizer nutrient. In 1863 Evan Pugh, a professor at the Pennsylvania Agricultural College, concluded that except for ashes, there was no economical source of potassium, although scientists had long known that it was important for plant growth and that it should be a constituent of a complete fertilizer.[70] Peruvian guano and fish waste contained small amounts of potassium, and ashes, especially from unleached hardwood, were good sources. But none of these materials could provide adequate supplies of potassium, Some of the early artificial guanos claimed to include potash, but commercial potash was too expensive for manufacturers to consider its use in a fertilizer. Fortunately, on most soils, potassium may have been the last of the three major fertilizing elements to be exhausted. Thus initially demand for potassium was not as great as for the other two elements.

In the late 1850s and 1860s entrepreneurs attempted to exploit New Jersey greensand marl commercially. They set up mining companies, extended rail lines to the marl beds, brought in mechanical diggers, and shipped the marl throughout the Northeast. Although the greensand marl contained significant amounts of potassium, it was too bulky and too slow-acting to succeed as a commercial fertilizer. Shipments exceeded 100,000 tons a year in the late 1860s, but afterward the industry declined rapidly and marl returned to its former status as a locally useful soil supplement.[71]

The potassium supply situation began to change in 1861 when deposits of potash salts were discovered near Strassfurt, Germany.[72] Imports into the United States did not start until 1873, when fertilizer dealers began to advertise kainite, a low-analysis potassium salt mined in Germany.[73] By 1874 the Pacific Guano Company was using 3 percent kainite in its "Soluble Pacific Guano," but the purpose was as much to preserve the fish scrap as to introduce potassium into the fertilizer.[74] Soon other companies also began including German potash salts in their products to produce a more completely balanced fertilizer.[75] Imports did not reach a significant scale until the 1890s, but from then until the outbreak of World War I Germany supplied most of this country's muriate of potash and other potash salts used in fertilizer. As the trade developed, it became a German monopoly, similar in many ways to the Peruvian guano monopoly half a century earlier.[76]

Although the South Carolina phosphates and German kainite gradually replaced recycled raw materials, the industry continued to seek out waste products which it could reuse. It continued to use bones and other slaughterhouse wastes on a large scale. Numerous plants sprang up in the South to process cottonseed into fertilizer.[77] H. J. Baker & Brother of New York developed and sold an effective fertilizer based on the refuse from their castor-oil business.[78] Sugarhouse wastes still found their way into fertilizers.[79] The Pacific Guano Company used refuse salt from Connecticut munitions plants to prevent its stock of fish scrap from spoiling.[80] In 1870 the *Country Gentleman* carried reports of a new superphosphate factory near Rochester, New York, which intended to use the refuse acid from kerosene oil refineries.[81]

Adulteration and Fraud

The ease with which fertilizers could be adulterated or fraudulently sold was one of the major factors that slowed the spread of commercial fertilizers. As the *American Farmer* warned in 1869, the fertilizer business provided "almost unlimited opportunity for fraudulent dealing."[82] Until ways were found to control these problems, farmers were reluctant to invest in fertilizer use and honest manufacturers found it difficult to sell their well-made products.

Dishonest dealers or manufacturers had been able to adulterate easily practically every fertilizer and soil supplement offered on the market. As early as 1797 farmers suspected that they were purchasing gypsum adulterated with ground chalk or limestone.[83] Even horse manure was adulterated. One Long Island farmer complained in 1827, after visiting the establishment of a New York manure dealer, that the product had been largely diluted by the addition of tanbark, sawdust, and other materials he considered useless.[84] Unscrupulous dealers added water, plaster, salt, chalk, and other adulterants to supposedly "pure" bone meal. They diluted unleached wood ashes with worthless anthracite coal ashes. Guano dealers mixed in large quantities of gypsum, limestone, brick dust, and other worthless substances. Sometimes the only genuine part of their product was the smell of real guano coming from the bags they were reusing.[85]

Superphosphates provided even better opportunities for fraud. Since farmers had little idea of what went on inside a fertilizer factory, it was easy for get-rich-quick operators to produce a superphosphate composed mainly of street dirt, lime, or swamp muck. In 1857 the *American Agriculturist* carried a cartoon (in figure 8-5) that summarized what many farmers must have thought about the integrity of superphosphate manufacturers. In this cartoon, laborers are busy carting the huge piles of swamp muck, red earth, and other worthless ingredients into the factory while the small amounts of sulfuric acid, guano, and bones, the only valuable ingredients in the drawing, remain untouched. Apparently the opportunities for fraud were limited only by the imagination of producers and dealers. The situation was so bad that in 1868 the *Boston Journal of Chemistry* could find no brand of genuine superphosphate to recommend.[86]

In addition to adulteration, several other kinds of fraud were also possible. Sometimes dealers merely switched labels and sold inferior varieties of guano as "No. 1 Peruvian." Another tactic was for a company to make up a special batch of high-quality fertilizer for analysis or experiments. For instance, the Lodi Company allegedly sent excellent samples of their poudrette to Daniel Webster and Andrew Jackson Downing, and

Fig. 8-5. Cartoon critical of superphosphate manufacturers. (From *American Agriculturist* 16 [1857]: 12.)

then used testimonials from these eminent gentlemen to promote an inferior product. Since many farmers were willing to test samples gratuitously sent by the companies, and many chemists were willing to provide favorable analyses for a fee, this deception was relatively easy.[87] As a result, many farmers assumed that none of the statements, testimonials, or analyses included in promotional literature were credible. The widely respected agricultural writer, Joseph Harris, warned that testimonials for fertilizers were "about as reliable as testimonials to a patent medicine," or in other words, "pre-eminently unreliable."[88]

The "Grafton Mineral Fertilizer & Destroyer of Insects" represents an extreme example of the kind of humbug which potential fertilizer purchasers encountered.[89] The Littleton, New Hampshire, entrepreneurs who introduced this fertilizer about 1867 claimed that it promoted the growth of "all vegetation" and that it was "death to canker worms, grubs, . . . and all other plant-destroying insects." Their advertising pamphlet prominently pointed out that its principal ingredient, "carbonic acid," was essential to plant life. It neglected to mention that scientists knew carbon dioxide was freely available for plants from the atmosphere. The pamphlet also noted the value of the lime, magnesia, and iron contained in the fertilizer without mentioning that these were available much more cheaply from commercial agricultural limes or were already available in most soils. Despite the bold advertising and twenty-seven pages of testimonials from satisfied New England farmers, Peter Collier, the secretary of the Vermont Board of Agriculture, doubted that the material was worth one dollar per ton.[90] He had an even lower opinion of the "Stevens' Mineral Fertilizer" mined one mile north of the Grafton quarry. His analysis showed it contained over 97 percent quartz![91]

The problem of fraud and adulteration was further compounded because many manufacturers and dealers did not fully understand their business. Honest dealers sometimes unwittingly received adulterated products. Moreover, honest manufacturers often followed mistaken theories and produced inferior products.[92] Sometimes undetected variations in the raw materials caused similar problems. The time and expense of accurate chemical analysis prevented farmers from testing their purchases. Even well-made fertilizer often appeared unreliable because of variations in the conditions of farmers' experiments. Overall, the extensive fraudulent practices and the widespread suspicion of these practices poisoned the entire fertilizer industry.

The Elimination of Fraud

Farmers, editors, and honest producers proposed three different methods to solve these problems. The first requested the agricultural press, agricultural societies, and state governments to use their power to expose fraud so as to protect purchasers. Several times in the 1850s the press did exactly that, although editors generally were reluctant to name or identify specific products. Therefore, most warnings were general advice for farmers to be careful in their purchases and to be aware of the possible kinds of adulterations they might expect. Sometimes the press published hints on how to detect fraud, but these probably were not very accurate. Nevertheless, the glare of adverse publicity probably deterred some entrepreneurs from fraudulent activity; and the reluctance of the leading journals to open their advertising pages to unreliable merchants may have hindered the less scrupulous operators. Until such scrutiny became more intensive in the 1870s, however, fertilizer producers were generally free to do as they pleased.

Various state officials, agricultural societies, and colleges also attempted to use the glare of adverse publicity to discipline the fertilizer industry. James Higgins, who had been appointed as Maryland's state agricultural chemist in 1848, began a series of reports in 1851 which included analyses and assigned monetary values to various guanos and manufactured fertilizers.[93] This practice had been pioneered by British agricultural chemists. As early as 1852 David Steward, the chemist of the Maryland State Agricultural Society and professor of chemistry at St. John's College, Annapolis, was also assigning monetary values to fertilizers he tested for the society.[94] Exposure of fraud was probably Samuel W. Johnson's motive in 1853 when, as a student at the Yale Analytical Laboratory, he analyzed two of the leading superphosphates. It was undoubtedly the goal of the Connecticut State Agricultural Society when it commissioned Johnson in 1857 to analyze fertilizers sold in the state and to publish the results.[95] The analyses published in 1862 by Evan Pugh of the Pennsylvania Agricultural College were intended to have the same effect.[96]

Perhaps the most successful attempt by a nongovernmental organization to clean up the fertilizer business was the investigation launched in 1872 by the New York State Agricultural Society. The society appointed a committee to purchase samples from the leading guano dealers in New York City and to have the samples tested by an analytical chemist. The tests showed that six of the ten samples were contaminated with sand or brick dust and only worth half their selling price.[97] The matter might have ended here with little embarrassment to the dealers, some of whom were well established and supposedly reliable.[98] The representative of the Peruvian government, however, decided to take legal action against the firms who had allegedly adulterated "pure" Peruvian guano.[99] During the ensuing court case, one of the prosecuted dealers, George Ricardo, acknowledged that he sold mixed guanos under the counterfeited labels of genuine Peruvian guano. When confronted with the evidence in his own books that he had purchased large quantities of molder's sand,

Ricardo also admitted that he mixed this with Jarvis guano before mixing the latter with Peruvian guano. He claimed that nine-tenths of his customers were guano dealers who knew his guano was adulterated, but specifically ordered the mixed guano from his factory. As a result of this testimony, the court, in a precedent-setting opinion, enjoined the accused dealers from selling anything which purported to be Peruvian guano.[100]

Government regulation, the second major solution suggested by reformers, was a logical outgrowth of the regulation of commodities such as drugs, flour, and tobacco common in many states. Moreover, as in so many other areas, American agricultural leaders looked to European precedents, especially in Great Britain and Germany, where fertilizer laws had already proven their effectiveness.

The first state to take direct action to regulate the fertilizer trade was Maryland, where the problem was most severe.[101] After considerable agitation in agricultural journals, the Maryland legislature passed a bill in 1847 requiring the appointment of a "practical chemist" to "inspect analytically" all guano sold within the city limits of Baltimore and imposing a $20 fine for anyone selling guano there without evidence of proper inspection.[102] The following year the legislature extended the protection of the act to the entire state.

The law was apparently not effective. James Higgins, Maryland state agricultural chemist from 1848 to 1858, criticized the inspection system severely in his early reports because the inspection grades used by the inspector bore no relation to the chemical content of the fertilizer and may have facilitated dealers in their attempts to deceive farmers.[103] Higgins's own analyses showed marked variations in the quality of guanos that the guano inspector considered "No. 1." As a result of considerable criticism, in 1854 the legislature strengthened the 1847 act so that the state inspector was now required to analyze the percentage of ammonia and phosphates in guano, make his analyses publicly available, and establish a grading system based on these analyses.[104] The law was apparently strong enough to drive the importers of Peruvian guano temporarily from Baltimore to New York, where they did not face any state inspectors.[105]

Nevertheless, Maryland's pioneering laws were probably limited in effect. As David Steward, the chemist of the Maryland State Agricultural Society, pointed out in 1857, unscrupulous dealers could easily misuse certificates issued by the state inspector to sell uninspected guano.[106] The act, moreover, applied only to guano. By the end of the decade, manipulated guanos and superphosphates were becoming more important commercial fertilizers, but the act did not apply to them. Not until 1868 did the Maryland legislature amend the inspection act so as to include manufactured fertilizers, and then the bill was inoperative because it failed to provide any fees to pay the inspector. An 1870 act removed this difficulty, but the act had to be further revised in 1886 and 1890.[107]

Meanwhile, other states were also attempting to regulate the fertilizer trade. Georgia passed a fertilizer control bill in 1868 and amended it the following year and again in 1874 to require the inspector to analyze fertilizers and publish the results.[108] The Maine legislature passed a bill in 1869 requiring all fertilizers sold in the state to carry labels specifying the maker or seller, and the percentages of ammonia, soluble phosphoric acid, and insoluble phosphoric acid.[109] The law imposed fines for failure to follow the labeling requirements and allowed farmers to sue sellers for failure to live up to the labeling requirements, but it was difficult to enforce. The same year, Connecticut passed a similar but stronger law, and Massachusetts passed a weaker one.[110] During the next decade, many of the other fertilizer-consuming states passed similar legislation. Delaware and Alabama passed laws in 1871, New Hampshire and South Carolina in 1872, New Jersey in 1874, Virginia and North Carolina in 1877, New York in 1878, and Pennsylvania and West Virginia in 1879. During the early 1880s, several more states passed such legislation, until by 1882, when the newly formed National Fertilizer Association published a manual for manufacturers summarizing state fertilizer regulations, twenty states had fertilizer contol legislation. These included every state along the eastern seaboard except Rhode Island and Florida and most of the states in the Ohio valley. These laws all required labels which included an analysis of the fertilizer. They also frequently provided for state inspection and fertilizer testing. They generally provided some form of fee or license tax to pay for the program, imposed fines on violators ranging as high as $200, and allowed purchasers to recover any damages in court caused by mislabeling. The first attempts in many of the states were ineffective, but legislatures often passed supplementary legislation to make the laws more effective.[111]

The impact of these laws is difficult to assess. Very few court cases seem to have been brought against dealers or manufacturers for failure to obey the law. When, in 1875, a Virginia judge ruled in favor of a farmer who refused to pay for a spurious fertilizer which violated the state's fertilizer control law, the editors of the *American Farmer* thought the event significant enough to report it in their pages as if it were the first such occurrence.[112]

The analyses published during the 1870s by the various state inspectors as required under some of the acts, and others published by state boards of agriculture and by independent institutions such as the Bussey Institution at Harvard, may have had a greater effect.[113] Sustained exposure to public scrutiny of this kind probably had a significant effect on manufacturers. The analyses themselves show that the quality of fertilizers improved significantly during the 1870s. In the case of Georgia, the state agriculture department's second report on fertilizers in 1875 to 1876 proudly noted there had been a 16 percent improvement in overall quality since the first report had been issued a year earlier.[114] Manufacturers and dealers whose products had shown up well in comparative fertilizer analyses advertised the results, possibly adding to the impact of the public analyses.[115]

A number of other factors may have contributed to the improvement in fertilizer quality, but there is some evidence that fertilizer manufacturers were bothered by adverse analysis. The best example is the case of George F. Wilson, who had founded the Rumford Chemical Works near Providence, Rhode Island, in partnership with Harvard's pioneering agricultural chemist, Eben H. Horsford, to produce "Horsford's Cream of Tartar Substitute."[116] In the late 1860s Wilson began producing "Wilson's Ammoniated Superphosphate," which was composed mainly of wastes and reject bones from baking powder production. Johnson analyzed this fertilizer and declared it worth only $18.71, although it sold for $65 per ton. In his 1873 advertising pamphlet, Wilson bitterly attacked Johnson's methods and state fertilizer inspection in general and strongly implied that the Connecticut *Homestead* was paid by a competitor for publishing material critical of his fertilizers. Nevertheless, he discreetly changed his formula, nearly doubling the amount of bone phosphates used in each ton from 65 pounds to 120 pounds.[117] Johnson subsequently found this fertilizer much improved.

Most of the impetus for fertilizer control legislation came from farmers and their advisers, although manufacturers also supported much of the legislation. Generally, enlightened agricultural reformers, such as Peter Collier in Vermont and W. O. Atwater in Connecticut, took the lead in the campaign for state legislation.[118] Farmers' clubs and state agricultural societies also sometimes initiated requests to legislatures for fertilizer regulation.[119] Agricultural editors generally supported these requests.[120] Manufacturers apparently objected only to a few laws they felt were especially burdensome because of the large lump-sum license fees required.[121] At least one manufacturer, Henry Bower of Philadelphia, actively supported the laws, and others may have realized how much they benefited from the market stability created by the laws.[122]

The third, and probably the most effective, line of attack on the problem came from the manufacturers. Agricultural editors and experts had long advised readers that the best safeguard against fraud was to purchase only from reputable and reliable dealers and manufacturers.[123] Manufacturers realized that fraud hurt their industry in general and that the only way to prosper was to establish the reputation of their company. The biographies of the early fertilizer manufacturers all discuss the importance of personal integrity in a business where the customer had so little else to rely on.[124] Some manufacturers supported reasonable state regulations as one way to shore up the public's faith in their products.[125]

Companies quickly resorted to brandnames such as "I Increase," "High Grade," "Soluble Pacific," "Tip Top," and "Patapsco." They also established trademarks to protect the integrity of their products. These trademarks were often highly individual and graphic, such as Baugh & Sons' cow with half its body cut away to reveal the skeleton, or the Manhattan Manufacturing and Fertilizing Company's heart dripping blood, skull and crossbones, and quotation from Lord Byron's *Childe Harold's Pilgrimage.* (See figure 8-6.) Advertisments often warned purchasers to beware of imitators and only buy products with the trademark.

PERUVIAN GUANO SUBSTITUTE.

BAUGH & SONS,

NONE GENUINE  ON EVERY

WITHOUT THE SACK AND BARREL.

Fig. 8-6. Fertilizer company trademarks: Manhattan Manufacturing and Fertilizing Company trademark for "Phosphatic Blood Guano" (from *Country Gentleman* 36 [1871]: 653); trademark of L.L. Crocker Company, Buffalo, N.Y. (from *Country Gentleman* 41 [1876]: 239); Baugh & Sons trademark (from Edwin T. Freedley, *Philadelphia and Its Manufacturers in 1867* [Philadelphia: Edward Young, 1867], 289).

The very names of some products, such as "Baugh's Pure Raw Bone Superphosphate," were designed to highlight their reliability. Advertising literature also stressed as graphically as possible the purity of the ingredients used in fertilizers. See, for example, the bone-covered advertisements for Baugh & Sons in figure 8-7 which dramatically conveyed the message that its fertilizers contained real bone. The prehistoric creatures depicted in the advertisements of the Bradley Fertilizer Company (figure 8-8) and the Marine and River Company were also designed to convey the message that the South Carolina rock phosphates used in their fertilizers were organic in their origin, even if the animals were prehistoric.

Probably the best way to establish a reputation, however, was to put out a reliable product, as the leading companies attempted to do. Naturally, those fertilizers successful in field trials and chemist analyses generally prospered, while adulterated or worthless products seldom survived more than a season or two. Once a farmer was satisfied with a brand, however, he tended to stick with it rather than experiment with new, unknown products on the assumption that brand loyalty would protect him from possible humbugs.[126]

In addition, some manufacturers voluntarily guaranteed the analyses on their bags, as manufacturers had done in England with good results. In the 1860s reformers such as Evan Pugh in Pennsylvania and Stephen L. Goodale in Maine had urged companies not only to print analyses on their bags but also to guarantee those analyses.[127] B. R. Croasdale &

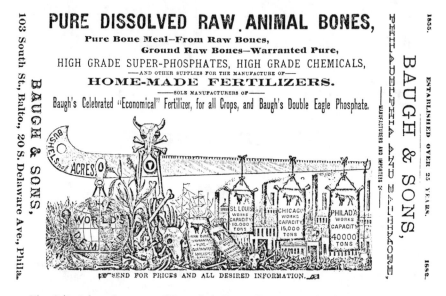

Fig. 8-7. Advertisement of Baugh & Sons. (From *Maryland Directory* [Baltimore: J. Frank Lewis, 1882], 86.)

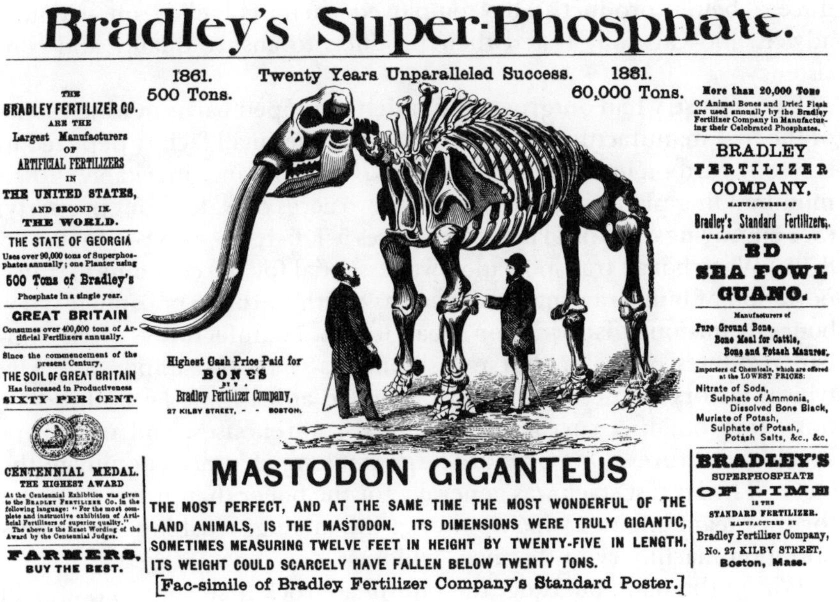

Fig. 8-8. "Mastodon giganteus" from standard poster of the Bradley Fertilizer Company. (From 1881 advertising pamphlet.)

Company of Philadelphia did this in 1867.[128] Other companies followed suit.[129] In Connecticut, dealers and manufacturers signed guarantee agreements drawn up by the state experiment station in 1876.[130] It is unclear just what legal remedy an aggrieved customer would have had, but the guarantees did add an extra level of reassurance.

Manufacturing Technologies

The third major source of instability within the fertilizer system was the poor and inconsistent quality of most manufactured fertilizers. Some of this was attributable to willful adulteration, but much was unintentional. Manufacturers did not understand or take into account the variable nature of their raw materials, did not have adequate manufacturing procedures to insure a consistent product, did not have adequate quality control measures, and often did not fully understand the principles of their processes. Unfortunately, until the companies turned out a better product, no amount of advertising could convince farmers to take it up.

By the 1880s manufacturers had put their plants on sounder technological footing. They were able to grind the ingredients much finer. They replaced hand mixing with more uniform mechanical mixing, and used power-driven machinery to facilitate most factory operations and to produce a better product. They employed chemists, relied on scientific advice, and carefully tested their products to ensure quality and consistency.

The industry had outgrown the rudely equipped barns in which some of the first manufacturers operated, such as the small facility depicted in the 1857 cartoon in figure 8-5. In the 1870s, engravings inevitably show a multibuilding plant situated on one or two acres of waterfront property. (See engravings of typical plants in figures 5-1, 6-1, 8-3, 8-4, 8-9, 8-10, and 8-11.) Waterborne transportation was essential for the economical transportation of bulky raw materials such as South Carolina phosphates, raw bones, brimstone, fish scrap or urban wastes. Manufacturers often shipped products by water, but rail shipments were also important. The typical factory in these engravings included at least one large three- or four-story building surrounded by sheds, warehouses, and numerous smaller structures. One or more large smoking chimneys dominate the scene. The largest stack was generally for the boiler that provided steam for the steam engines which powered the mixing, crushing, and transportation machinery in the main building.

Within the main building, the phosphate rock first went to powerful iron crushers. Then a spiral elevator conveyed the crushed rock to a drying machine, after which another spiral elevator conveyed it to pulverizing machines and sets of iron rollers that reduced it to a fine powder. The material was then bolted and any coarse particles returned to the mills to be reground. Next, elevators took the ground phosphate to storage bins on the top story. From there it flowed to a Poole and Hunt

Fig. 8-9. Works of the Rasin Fertilizer Company, near Baltimore, Maryland, 1882. (From *Maryland Directory* [Baltimore: J. Frank Lewis, 1882].)

Fig. 8-10. Lister Brothers plant in Newark, New Jersey, 1882. (From *Maryland Directory* [Baltimore: J. Frank Lewis, 1882].)

mixing machine, which could handle about one ton at a time. (See figure 8-12.) Measured amounts of phosphate rock, sulfuric acid from a reservoir on the top floor, and ammoniating ingredients were added as the mixer rapidly revolved. After being thoroughly mixed for about fifteen minutes, the pastry mixture was run out onto the floor below. There it was allowed to set up for a day or two as the chemical reactions were completed. Laborers then broke it up by pickax, passed it through a disintegrator on the second story to reduce it again to a powder, and finally bagged it. Most of the larger plants had acid chambers on the premises to produce sulfuric acid from brimstone or pyrites. After bagging, the superphosphate was sent to a warehouse on the property to await shipment, usually between late fall and early spring. Numerous smaller support structures were either attached to the main building or nearby. In addition to warehouses for the storage of various raw materials, the typical plant had a boiler house, scale house, firepump house, office, blacksmith shop, stables, and sometimes residences for workmen.[131] For a diagram of a typical plant, see the 1880 insurance map of the Pacific Guano Company's Charleston works in figure 8-13.

The highly mechanized internal transportation systems resembled those of contemporary flour mills, but the actual production was a batch process, instead of a continuous process as in the flour mills, and considerable hand labor was included, especially in emptying the dens in which the superphosphate hardened. Although manufacturers made considerable improvements in their plants during the 1870s and 1880s, the production

Fig. 8-11. Woods Hole Plant of the Pacific Guano Company, about 1877. The main factory building is immediately behind the tallest chimney, with the top masts of a ship showing behind it. The first floor was for crushing and grinding, the second for packing and shipping, the third for mixing and screening. The boiler house is a separate building to the right of the chimney and in front of the main factory. The small building to the right of the road is the company office. Behind the office and to the right is a frame storage building. The flat-roofed building in the distant right is the acid chamber building and attached furnaces, boiler room, and storage sheds. An underground pipeline connected the acid building with the main plant. The small building in front of the chamber building may be the company stable. Compare this view with earlier view in figure 5-1. Despite the heavy wagon traffic shown on the road leading to the plant, all raw materials and almost all the finished fertilizers were moved by water. (Drawing from G. Brown Goode, "A History of the Menhaden," U.S. Commission of Fish and Fisheries, *Report of the Commissioner for 1877* [Washington: GPO, 1979]. Building names and functions are from 1880 insurance map in Pacific Guano Company Collection in the Baker Library. The artist, however, seems to have taken liberty with the perspectives.)

process remained a batch process.[132] Generally, the size of the batch was determined by the size of the mixers and of the dens in which the fertilizer was cured. The only important innovation was the development of mechanical means of emptying the bins.

The resemblance between fertilizer mills and flour mills was probably more than coincidental. The same firms that supplied equipment for flour mills also supplied the fertilizer mills. For instance, the Baltimore machinery manufactory of Poole and Hunt, which supplied many fertilizer plants with crushers, mixers, and other machinery, also produced

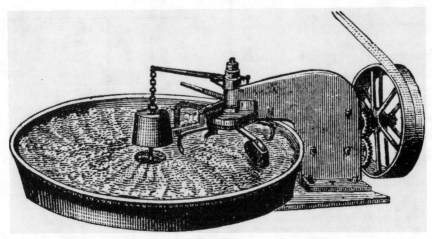

Fig. 8-12. Poole and Hunt mixer. This machine, designed by the Baltimore firm of the same name, was the first successful mechanized fertilizer mixer. The large iron pan revolved horizontally around a support. The set of plows revolved on its support. There is also a guiding device to direct the finished product towards the central opening in the pan, thorugh which the semi-liquid material dropped through a conduit to the floor below. The chainheld object in the center may be a stopper for the exit hole. (Engraving from Campbell Morfit, *A Practical Treatise on Pure Fertilizers* [New York: D. Van Nostrand, 1872], 145.)

flour-milling machinery.[133] Similarly, the Boston firm of Holmes & Blanchard, which supplied "disintegrators" and other machinery to the Pacific Guano Company, also sold millstones and equipment for flour mills.[134]

The introduction of practical chemistry into the factory was probably more important than the development of efficient, mechanized plants. Chemists understood the variations in raw materials and the tendency of impurities to absorb sulfuric acid or produce undesirable reaction products. They could determine precisely the amount of acid needed to acidify completely a given volume of rock without leaving free sulfuric acid that made the product hard to handle. Combined with the fine grinding of the raw materials, the accurate regulation of ingredients made it possible for manufacturers to produce a much better product.

By the 1870s most of the larger fertilizer companies had chemists. Often the chemists were partners in the firms, as was the European-born and -trained Dr. Gustav A. Liebig in the Baltimore firm of Liebig and Gibbons.[135] Sometimes the company chemist was also superintendent of the plant, as in the case of N. A. Pratt in the Charleston Mining and Manufacturing Company. Other chemists were only consultants, such as L. R. Gibbes, a professor at the College of Charleston who served as chemist for the Stono Phosphate Company at Charleston, or St. Julien

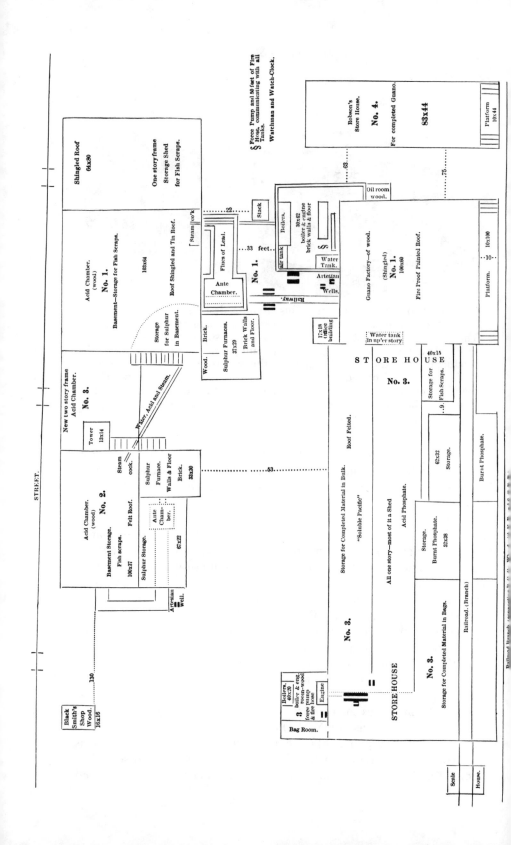

Ravenel who not only helped organize the Wando Company, but also served as chemist for the Pacific Guano Company's Charleston plant when it was set up. Charles U. Shepard, Jr., son of the discoverer of the phosphatic component of the local marl, performed a similar service for J. B. Sardy's Wappo Mills. Companies often considered the services of a chemist important enough to feature it prominently in their advertising.[136]

The best documented of these chemists is Azariah Foster Crowell, son of Prince S. Crowell, one of the founders of the Pacific Guano Company. Although he had no formal education in chemistry, he served competently as chemist and superintendent of the Woods Hole plant during most of the 1870s and 1880s. His 1884–1885 formula book contains one hundred different formulae to make fertilizers from a variety of raw materials and for several different markets.[137] The surviving correspondence of the company includes requests to Crowell from the manager of the South Carolina works for a new formula for "Soluble Pacific Guano" after he was forced to change raw materials.[138] Crowell's notebooks show that he carefully tested the product to make sure that the company's standards were maintained.[139] He also carried out experiments to discover better, more efficient ways to produce fertilizers. Chemists apparently performed similar functions for many of the other companies by 1875, when the editors of the *American Farmer* noted that commercial fertilizers were prepared "by the aid of scientific skill" that a few years earlier had been "entirely unknown in their manufacture."[140]

Overall, the introduction of chemical expertise and the development of effective, mechanized grinding and mixing machines enabled fertilizer manufacturers to produce the consistent, reliable product which was essential for long-term acceptance of commercial fertilizers by farmers. Combined with the solutions to the severe raw material problems and the gradual elimination of fraud and adulteration, these manufacturing improvements eliminated the most severe barriers to the development of the fertilizer industry, allowing the manufacture of fertilizers to grow rapidly and the use of fertilizers to spread into most areas where soil conditions and economic circumstances made them rational.

Fig. 8-13 (opposite). 1880 insurance map of Pacific Guano Company works in Charleston, South Carolina. For insurance purposes, the company valued this plant at $89,000. (From Pacific Guano Company Collection, Case 1, Baker Library, Harvard University Graduate School of Business Administration.)

CHAPTER IX

The Maturation
of the Industry

With its problems of raw material, production, and public acceptance largely solved, the industry was able to develop rapidly to meet the growing demands for chemical fertilizers. According to federal census data (which is not entirely reliable) summarized in table 9-1, the number of fertilizer plants grew from 47 in 1859 to 126 in 1869, 364 in 1879, 390 in 1889, and 478 by 1899.[1] During this period, fertilizer plants became much larger and more capital intensive. In 1859 the average capital per plant was less than $10,000, but grew to almost $50,000 in 1879 and approached $150,000 by the end of the century. Total production increased from less than $50,000 in 1859 to $23,000,000 in 1879 and $44,000,000 in 1899. The actual tonnage of nutrients produced probably increased even more rapidly since the price of fertilizer decreased dramatically during most of the period. In the North the price of ammoniated superphosphates, the most common fertilizer, declined from $55 per ton in 1868 to $50 in 1872, $40 in 1879, $35 in 1889, and $33 in 1894.[2]

The Industry Shifts Southward

As the industry developed, it gradually assumed its modern geographical configuration with manufacturing located as close as possible to markets and phosphatic raw materials. Economic considerations encouraged the industry to shift southward where both the principal markets and phosphate deposits were located in the late nineteenth century, but historical factors, including the continuing connections with the urban recycling system, exerted a strong influence over keeping the location of fertilizer manufacturing in the North.

In 1879 the seaboard states from Maryland north to Massachusetts still produced over three-fourths of the nation's fertilizers, but consumed only 44 percent of the product. Georgia, Alabama, and the Carolinas consumed 36 percent of the fertilizer used in the nation, but produced only 15 percent, even though the principal raw material came from the region.[3] The typical manufacturer imported South Carolina phosphates to a northern city, manufactured them into fertilizers and then shipped the finished product back to the Southeast. When the Georgia Department of Agriculture analyzed 101 brands of fertilizers sold in Georgia during 1875–1876, of the forty-eight companies represented, only seven were based in Georgia, eight manufactured their fertilizers elsewhere in the South (mainly at Charleston), sixteen were from the Chesapeake region (mainly in Baltimore), and seventeen were from farther north.[4]

The heavy concentration of fertilizer firms in northern cities in the 1870s was probably illogical economically. Because fertilizers are high in bulk per value, transportation costs constituted an important part of the retail price. Moreover, the finished superphosphate weighed approximately twice as much as the raw rock phosphate. Therefore, companies located near their markets had an economic advantage that should have encouraged more of the industry's manufacturing capacity to move to the South, as it eventually did by the end of the century.

The rearrangement of the industry to conform to economic rationality after the development of South Carolina phosphate deposits and the rise of the southern demand for commercial fertilizers may have been delayed because of the high profit margins in the fertilizer industry of the 1870s and 1880s. Shipping rates for fertilizers between the North and South were generally less than $2 per ton.[5] Since manufacturing costs for an efficient northern operator such as the Pacific Guano Company ranged between $18 and $22 per ton, and selling prices were generally above $45 per ton, the transportation costs initially may not have been a significant barrier to the profitability of northern plants.[6] Moreover, the South's higher fertilizer prices, about 20 percent above those of the North, may have minimized the impact of the extra transportation costs incurred in shipping raw materials and finished products back to the South.

Several other factors contributed to the continuing preponderance of northern manufacturers. When demand for fertilizer first developed in the South, the industry was already well established in the North where it had grown out of the urban-rural recycling system. Northern plants also had more ready access to supplies of fish scrap and slaughterhouse refuse, used to nitrogenate fertilizers. In addition, the southern fertilizer industry faced the same scarcity of capital and manufacturing expertise which plagued many other nascent southern industries.

By 1900 the disparity between northern and southern production was narrowed as capital, expertise, and manufacturing capacity gradually shook off their connections with the urban recycling system and moved nearer the principal markets and raw materials in the South.[7] As fertilizer

Year	Number of establishments	Total capital	Value of products	Fertilizer expenditures by farmers	Capital per plant	Production per plant	Employees per plant
1859	47	466,000	891,344	—	9,914	18,957	—
1869	126	4,395,948	5,815,114	—	34,880	46,000	16
1879	364	17,913,660	23,650,795	28,586,397	49,213	64,972	23
1889	390	40,594,168	39,180,844	38,469,598	104,087	100,463	26
1899	422	60,685,753	44,657,385	54,783,757	143,805	105,823	27

Table 9-1. Fertilizer statistics. (From U.S. Census Office, Twelfth Census [1900], vol. 5, *Agriculture*, pt. 1 [Washington: GPO, 1902], cxi.)

prices fell during the last third of the century, transportation costs exerted a greater impact on decreasing profits, but northern plants continued to manufacture a disproportionate part of the national total.

Urban Centers of the Industry

Throughout the nineteenth-century development of the fertilizer industry, Baltimore remained its center.[8] Not only did Maryland produce more fertilizer than any other single state, but Baltimore manufacturers took the lead in developing the industry. Baltimore machine firms supplied equipment to fertilizer plants throughout the country.[9] Baltimore manufacturers were instrumental in forming the first national fertilizer manufacturers' associations in the late 1870s and early 1880s.[10] Most of the fertilizer companies further north established agencies or branches there.[11] Baltimore supplied large parts of the South and some of the Midwest. Baltimore merchants often shared in the financing and management of new plants as the industry spread into new areas.[12]

Within the North, most of the major plants were in or near Baltimore, Philadelphia, New York, and Boston, although Wilmington, Providence, and a few other smaller cities developed some manufacturing capacity. Many of these companies had strong historical connections with the urban-rural recycling system. Some had developed from recycling operations that processed bones, slaughterhouse refuse, and other urban wastes into fertilizers. Others had begun in the guano trade that replaced the recycling system in some areas. Regardless of their origin, most northern urban fertilizer plants continued to use recycled wastes to supply nitrogen for their mixed fertilizers, even though they had substituted South Carolina rock phosphates and German potash salts for guano or bone as sources of potassium and phosphorus. As the industry expanded into the Midwest, it became part of similar recycling systems associated with the major slaughtering centers of Chicago, Cincinnati, St. Louis, and Buffalo. Urban locations offered several additional benefits to fertilizer companies besides the connections with the recycling system. Cities provided easy access to transportation, market mechanisms, financing, and technological expertise. The only nonurban production facilities in the North were those associated with the fish guano industry, such as the Quinnipiac Company near New London, the Pacific Guano Company in Woods Hole, and the Cumberland Bone Company in Boothbay, Maine.

Within each urban area the plants tended to locate on the outskirts, away from heavily populated areas. Around New York, the largest operations were across the Hudson in New Jersey, on Barren Island in Jamaica Bay, and in remote areas of Staten Island and Brooklyn. Most of the fertilizer factories near Baltimore were in the Federal Hill district immediately south of the city center, or farther south at Canton.

Of course, the main reason for these remote locations was the problem of odor which had plagued the industry since its earliest days. In addition to the pungent smell released by the storage and processing of fish scrap, slaughterhouse offal, and other organic wastes, the treatment of phosphate rock with sulfuric acid produced noxious fumes. Firms attempted to minimize some of the odor problems, but were not very successful. In Boston, a hostile board of health even temporarily closed the inappropriately named "Inodorous Rendering and Superphosphate Works" because of the plant's stench.[13]

Even though many of the newer fertilizer plants were not directly connected with the northern urban-rural recycling system, they had historical connections with that system because as the industry expanded geographically, northern fertilizer companies established branch manufacturing operations in the new areas. The Pacific Guano Company built a Charleston, South Carolina, plant, as did several other northern companies; the Darling Company of Pawtucket had works in Chicago by 1882; Baugh & Sons of Philadelphia had plants in Chicago by 1867 and St. Louis by 1882; and William Bradley of Boston had plants in Baltimore, South Carolina, and even California by the 1890s.[14]

Improvement in the Product

As the industry matured, its products became more suitable for sustained agricultural use. Samuel Johnson's analyses in the 1850s and Evan Pugh's 1862 analyses had revealed wide variations in quality between brands and between samples of the same brand tested in different years. From 1869 until the end of the century, however, the leading ammoniated superphosphates tested by Johnson and his successors in Connecticut remained remarkably constant in quality. They generally had 2 or 3 percent nitrogen, 6 to 8 percent soluble phosphoric acid, and up to 2 percent potash.[15] Other testing programs showed similar results.[16] These fertilizers were reasonably well-balanced complete fertilizers, although nutrient concentrations ranging from 8 to 13 percent are not high compared to modern fertilizers, which often have several times as much. The most important factor revealed by these analyses was their consistency, without which farmers would never have been able to rely on commercial fertilizers.

Another important development, which was just beginning in the 1870s, was the production of fertilizers specifically formulated for particular crops, such as tobacco, cotton, wheat, or corn. The industry had made a false start in this direction earlier in the century when some scientists recommended that the percentage of nutrients in a fertilizer should correspond to the percentage of the same minerals in the ash of the plants, but this approach proved to be erroneous. Attempts to formulate fertilizers based solely on soil analyses had fared little better, although in principle the amount of fertilizer needed should be directly related to the amount of nutrients in the soil and the amount a crop would remove. In the 1870s and 1880s scientists still used soil and plant analyses, but relied more heavily on information gleaned from careful plat experiments with crops under field conditions, and fertilizer companies began to consider scientists' carefully researched recommendations in formulating fertilizers. The "Stockbridge Manures" developed by Prof. Levi Stockbridge of the Massachusetts Agricultural College and introduced commercially by W. H. Bowker of Boston in 1876 are a good example of this trend.[17] It was probably not coincidental that Bowker, a leader in this movement, was himself one of the early graduates of the Massachusetts Agricultural College. Other companies such as Mapes in New York and Bradley in Boston offered similar fertilizers.[18] These products were probably closer to the actual needs of crops than earlier "complete fertilizers," although independent analyses sometimes showed a remarkable similarity between fertilizers supposedly formulated for different crops.[19]

By 1880 the fertilizer industry offered farmers a bewildering variety of products capable of meeting most of their needs and ingrained prejudices. In 1875 the Georgia Department of Agriculture tested 102 brands of fertilizers. Two years later the Connecticut Agricultural Experiment Station tested 150 different fertilizers. In 1892 the Virginia State Board of Agriculture analyzed an astonishing 380 different brands of fertilizers, including 275 varieties of ammoniated superphosphates.[20] In addition to brands of "standard" products, generally superphosphates or ammoniated superphosphates, and an array of special mixtures, many companies offered single ingredients such as sulfate of ammonia and muriate of potash.[21]

Almost every successful fertilizer introduced during the century was still available in the last quarter of the century. Peruvian guano could be had, although the Chincha Islands had been largely exhausted.[22] Unleached wood ashes, generally from Canada, were still available from dealers in the northern states.[23] Many manufacturers offered raw bone meal, or superphosphates made from it, long after the introduction of rock phosphates.[24] Navassa guano was still imported until almost the end of the century. Fish guano was still produced. A brisk trade in city stable manure continued as long as horses provided the prime motive power for urban transit systems.[25] Hence, farmers faced a wide choice. They could continue to use old, reliable favorites and supposedly avoid the risks associated with the newer manufactured fertilizers; they could buy ingredients and mix their own; or they could purchase some of the newer complete or special fertilizers.

The Use of Commercial Fertilizers

The use of commercial fertilizers had become general enough by 1880 for the Census Office to take a statistical interest in the subject. In that year's census, the enumerators asked farmers to report the value of commercial fertilizers they purchased. Even though the data is not as accurate as one might wish, it provides the best picture of the patterns of fertilizer use at the time. As the map in figure 9-1 and table 9-2 show, use of fertilizers was confined to the eastern seaboard, except for a few isolated counties in the Midwest and along the Gulf Coast. The original thirteen states accounted for 89 percent of the total fertilizer purchased. In the Northeast, which consumed about 33 percent of the total, the heaviest usage was around the major cities—Boston, Providence, New York, and Philadelphia. In the border states, which used about 19 percent of the total, the area surrounding Chesapeake Bay was the biggest user. Here again the urban influence is apparent since fertilizer use centered around Baltimore and the market gardening areas near Norfolk.

The transition toward general acceptance of superphosphates and other manufactured fertilizers took place in many areas along the eastern seaboard during the 1870s. By the middle of the decade many farmers described the use of these fertilizers as "somewhat general" or "extensive" in their vicinity. In 1873 the Massachusetts State Board of Agriculture's fifth report on fertilizers noted the "rapid progress among farmers" in the use of commercial fertilizers as part of a "rational system of manuring farmlands."[26] In 1875 the editors of the *American Farmer* predicted the nation was entering an age when the use of commercial fertilizers would be "well-nigh universal."[27]

A statistical analysis of the 1880 manuscript agricultural census returns for Riverhead, New York, demonstrates how widespread the use of fertilizers was in some areas.[28] This eastern Long Island town was well beyond the market gardening belt around New York City, but its agriculture was still closely connected with the urban markets. Probably the fertilizer practices there were similar to those of the other northeastern areas that averaged $1.00 or more in gross cost of fertilizers purchased per acre (shaded dark on map in figure 9-1). According to data in the manuscript agricultural census returns summarized in table 9-3, 197 of the 233 operating farms in Riverhead were using purchased fertilizers in 1879. On an average, these 197 farms spent $99.59 on fertilizer, or $2.31 for each acre they tilled.[29] This represented 13.3 percent of the total value of their produce and was enough to provide minimal applications on each tilled field every other year. The 197 farms using purchased fertilizers were roughly similar to the farms not using such fertilizers. The most significant difference was that the farmers who used purchased fertilizers

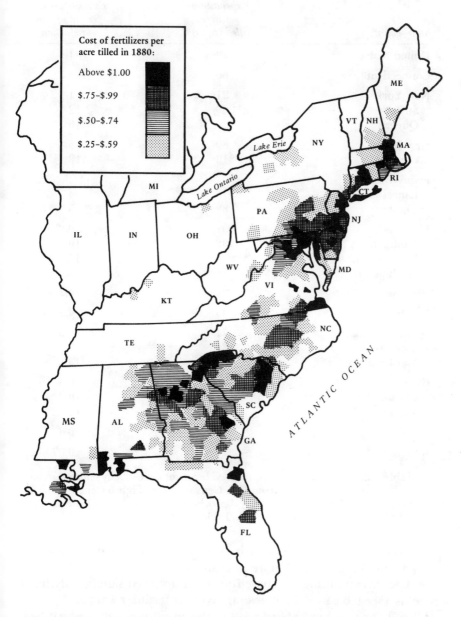

Fig. 9-1. Map of eastern United States showing intensity of fertilizer usage in 1880. Data accreted by county. (Calculated from U.S. Census Office, Tenth Census [1880], vol. 3, *Report on the Productions of Agriculture* [Washington: GPO, 1883], 104–40. Map adapted from U.S. Department of Interior, Geological Survey, *National Atlas of the United States* [Washington: GPO, 1970], 285.)

State–county	Expenditure per acre tilled (cents)	Expenditure as a percent of farm production
United States	12	1.3
Connecticut	66	2.7
New York	21	1.5
Suffolk	209	12
Queens	530	16
New Jersey	89	5.4
Camden	316	12
Pennsylvania		
Lancaster	75	4
Chester	88	5
Maryland	91	9.8
Montgomery	182	19
Howard	224	17
Washington	75	6.4
Virginia	29	4.7
North Carolina	35	4.1
South Carolina	71	6.5
Marion	131	10
Marlborough	225	13
Georgia	56	6.5
Clayton	117	10
Walton	124	11
Ohio	3	0.4
Iowa	0.5	0.07

Table 9-2. Expenditures for purchased fertilizers in 1880, by selected states and counties. (Calculated from U.S. Census Office, Eighth Census [1880], Vol. 3, *Report on the Productions of Agriculture as Returned at the Tenth Census* [1880] [Washington: GPO, 1883], Table VI, 102–40.)

devoted somewhat more of their cropland to wheat and hay, and generally farmed their land more intensively.

The farmers using purchased fertilizer achieved significantly higher yields. (See table 9-4.) Their average cost for fertilizer was $2.74 per acre tilled, but since they applied most of this fertilizer to corn, wheat, potatoes, and market garden crops, which occupied less than two-fifths of the tilled acres, the average cost of fertilizers for these crops must have been nearly $7.00 per acre. This would have purchased between 200 and 350 pounds of typical concentrated fertilizers, about what many agricultural experts recommended.[30] These liberal applications of fertilizers may have been an important factor in the larger yields achieved by these farmers. They reported 25 percent more corn per acre, 56 percent more wheat per acre, 61 percent more potatoes per acre, and 51 percent more hay per acre.

An analysis of a sample of Maryland farmers, summarized in table 9-5, confirms the findings of the Riverhead study. Of 74 randomly selected farmers in Washington County, which is located on the western fringe of the area of intensive fertilizer use in Maryland, only 16 purchased no fertilizers. Although the average farmer used only $.86 worth of fertilizer per acre tilled compared with $2.31 for Riverhead, nevertheless the use of fertilizer had become the general practice in the area. As table 9-5 shows, the farmers using fertilizers were generally similar to those not using fertilizers, except that they devoted a significantly larger portion of their land to their principal cash crop, wheat. They achieved slightly higher per acre yields, but their farms produced about the same values of products per acre as those not using fertilizers.

Farmers in the Chesapeake area had probably made the transition to purchased fertilizers even more completely than those on Long Island. As the 1880 federal census indicated, the farmers in the eastern part of Maryland and Virginia used commercial fertilizers extensively. Montgomery County, Maryland, which was now connected directly to metropolitan Baltimore by railroad, reportedly spent over one-half million dollars annually for fertilizers by 1879.[31] In 1873 a farmer from Sandy Spring, where the use of commercial fertilizers had first taken hold in Montgomery County, claimed that the use of guano and phosphates had been "universal" there for twenty-five years.[32] The manuscript agricultural census returns for the enumeration district which included Sandy Spring reveal that the use of commercial fertilizers was indeed almost "universal" by 1879. Of the seventy-eight farmers in the district, only one did not report the purchase of any fertilizers. The same year, the *American Farmer* reported a debate at the Deer Creek Farmers' Club in Harford County, Maryland, in which most of the members of the club thought it better to go into debt than to do without purchasing commercial fertilizers.[33] In Virginia by 1880, 70 to 80 percent of the tobacco crops in some areas were fertilized by "special manures" and farmers in areas along the North Carolina border which had been covered for years with dwarf oaks, brown sedge, and pines, transformed their wasteland into prime tobacco fields with the help of commercial fertilizers.[34] Clearly, many farmers in this region no longer questioned whether they should use commercial fertilizers, but only how much or what kind they should use.

Perhaps the transition from suspicion to acceptance of commercial fertilizers can be traced most clearly in the pages of the *American Agriculturist*. This popular agricultural journal had been suspicious and critical

Land Usage	Farms purchasing fertilizer (a)	Farms not purchasing fertilizer (b)	Difference (a − b = c)	Percent of difference (c/b)	2-tail probability†
Number of farms (total 233)*	197	36	—	—	—
Acres tilled	39.8	30.8	9	29%	0.063
Value ($) per acre tilled (including improvements)	99	115	−16	−14%	0.499
Acres in wheat	3.7	1.5	2.2	146%	0.000
Mean of wheat acreage as a percentage of total acres tilled	8.6%	5.3%	3.3%	62%	0.001
Acres in potatoes	1.9	1	0.9	90%	0.000
Mean of acres in potatoes as a percentage of total acres tilled	5.3%	4.6%	0.7%	15%	0.494

Acres of hay (excluding permanent meadow)	7.7	3.5	4.2	120%	0.000
Mean of acres in hay as a percentage of total acres tilled	17.8%	9.7%	8.1%	84%	0.000
Mean value of market garden produce as a percent of total value of produce	6.7%	4%	2.7%	67%	0.303
Livestock per acre tilled	0.19	0.17	0.02	12%	0.488
Acres in corn	6.2	4.1	2.1	51%	0.001
Mean of acres in corn as a percentage of total acres tilled	17.4%	17.2%	0.2%	0.1%	0.922

*The enumerators listed 247 farms in Riverhead, but 24 were eliminated from consideration here because they did not cultivate any field crops or market gardens. One of the two enumerators added footnotes to some of these cases indicating that the farms were part of an estate or that they were "not worked much" in 1879. Most of the other cases eliminated were under three acres. They reported no value of produce and probably were not actually farms, but large houselots.

†Two-tail probability of separate variance calculated in T-test.

Table 9-3. Comparison of farms using purchased fertilizers with those not using purchased fertilizers for Riverhead, New York, in 1880. (Calculated from 1880 agricultural census manuscript returns, manuscripts, New York State Library, Albany.)

Productivity	Farms using fertilizer (a)	Farms not using fertilizer (b)	Difference (a − b = c)	Percent difference (c/b)	Market value of difference*	2-tail probability†
Tons of hay per acre	1.03	.68	.35	51	$ 4.55– 7.00	0.001
Bushels of corn per acre	33.1	26.4	6.7	25	3.55– 4.02	0.005
Bushels of wheat per acre	14.4	9.2	5.2	56	5.20– 7.80	0.001
Bushels of potatoes per acre	103	64	39	61	12.48– 28.47	0.000
Value of product per acre	$20.60	$17.38	$ 3.22	14	3.22	0.147
Cost of fertilizers per acre	$ 2.74	—	$ 2.74	—	2.74	—

*Market values (New York wholesale prices calculated from *American Agriculturist* 36 [1877]: 165, and 38 [1879]: 371): Wheat $1.00–$1.50 per bushel, corn $.50– $.60 per bushel, hay $13.00–$20.00 per ton, potatoes $.32–$.73 per bushel. †Two-tail probability of separate variance estimate calculated in T-test.

Table 9-4. Comparison of results of use and non-use of fertilizers in Riverhead, New York, in 1880. (Calculated from 1880 agricultural census manuscript returns, manuscripts, New York State Library, Albany.)

Land Usage and Productivity	Farms purchasing fertilizers	Farms not purchasing fertilizers	2-tail probability*
Number of farms	58	16	—
Mean acres tilled	102.7	99.2	0.870
Mean value of produce per acre ($)	12.7	13.5	0.876
Horses and cattle per acre tilled	0.154	0.164	0.831
Yield of corn (bu.) per acre	34.8	29.6	0.154
Yield of wheat (bu.) per acre	15.7	13.8	0.338
Percentage of tilled land in wheat	35.7	28	0.100
Percentage of tilled land in corn	22.1	20.1	0.561

*Two-tail probability of separate variance estimate calculated in T-test.

Table 9-5. Comparison of farms using purchased fertilizers with farms not using purchased fertilizers for a randomly selected sample of farms in Washington County, Maryland. (Calculated from 1880 agricultural census manuscript returns, manuscripts, Maryland Historical Society, Baltimore.)

of most commercial fertilizers throughout the fifties and sixties. For example, in 1862 when the *Agriculturist* printed Evan Pugh's analyses of fifteen commercial fertilizers, Orange Judd, the paper's influential editor, appended his opinion that phosphatic fertilizers (including super-phosphates) were not essential for plants. He doubted their worth, regardless of the value Pugh assigned to phosphoric acid.[35] In 1870 the journal still barely mentioned commercial fertilizers. There were only a few insignificant articles and a few advertisements. Almost all of the articles dealing with soil fertility were concerned with improvements in methods of producing or storing farmyard manures, composts, or other locally obtainable organic material. During the spring fertilizer season, only six individuals placed a total of eleven advertisements. Only one of these, E. Frank Coe of Brooklyn, offered a superphosphate as his principal product. When asked what was the best substitute for stable manure, the editors responded that bone dust would be if it could be obtained.[36]

By the middle of the decade the attitude of the *Agriculturist* had changed dramatically. The number of articles on artificial or commercial fertilizers far exceeded those on barnyard manure or composts. In 1872 the editors began to counsel farmers not to be prejudiced against chemical manures since these were often made from farm products and helped make up for the loss in soil fertility caused by the sale of crops and animals.[37] Although the editors now accepted commercial fertilizers, they were still using the concepts of the recycling mentality. They still thought in terms of returning to the soil the nutrients which crops had removed, but now they were willing to consider returning nutrients from commercial sources instead of directly recycling nutrients.

The *Agriculturist* reflected the changed attitude in numerous other ways. Correspondents began to describe widespread use of superphosphates in their vicinity.[38] Joseph Harris, author of the "Walks and Talks on the Farm" series, began to recommend superphosphate and nitrate of soda.[39] Instead of suggesting that farmers produce their own superphosphates, the editors now acknowledged that "artificial fertilizers" were "so largely manufactured" that farmers could purchase them much more cheaply than they could manufacture their own.[40] The editors even advised that Charles V. Mapes, the son and successor of the notorious "Professor" James J. Mapes, was a trustworthy superphosphate manufacturer.[41] In deference to growing interest in commercial fertilizers, the "Market Report," which listed the price of various agricultural commodities in each month's *Agriculturist*, began listing prices of various commercial fertilizers in New York.[42] Even the index of the *Agriculturist* acknowledged the transition. Whereas previously it had listed the few articles it purchased on superphosphates and other commercial fertilizers under the heading "manures," it began to use the separate heading "fertilizers" for the more extensive articles it now published on the subject.[43] By 1877 the editors could proudly trace the progress made in commercial fertilizers and applaud the role their paper had played in the process.[44]

The *American Agriculturist* also mirrored the increased interest in science that accompanied the spread of chemical fertilizers. Since its inception the *Agriculturist*, like most other agricultural journals, had run occasional articles by leading American agricultural chemists such as Norton, Horsford, and Johnson and had reprinted scientific articles from British journals. The editors had always been skeptical about the usefulness of scientific information, but their attitude changed in the mid-1870s. The *Agriculturist* now carried a long series of articles entitled "Science Applied to Farming," by W. O. Atwater, Johnson's friend and colleague. Perhaps the reasons for the new attitude were best summed up in an 1875 article by Harvard botanist Asa Gray entitled "Practice and Science Agree." In this article Gray described how chemists had recently discovered that, as farmers had always known, leaf mold was indeed useful for growing plants despite the previous belief of many chemists that humus "did not amount to much" and could not supply plants with nitrogen.[45] Apparently agricultural science had finally advanced to the point where it could be of practical assistance to farmers.

Numerous other indicators pointed to the quickened interest of agriculturists in commercial fertilizers. Like the *American Agriculturist*, the *Country Gentleman*, which had previously not supported commercial fertilizers, began to pay considerably more attention to the subject in the mid-1870s.[46] In 1872 the United States Department of Agriculture, for the first time, sent out circulars in an attempt to obtain county-by-county information on fertilizer use.[47] The fertilizer statistics, which the New York state census had collected since 1855, suddenly became more accurate in 1875. Of course, the federal census followed suit in 1880 with its question on fertilizer purchases. The pressure exerted by farmers for fertilizer control laws, the interest in fertilizers shown by state boards of agriculture, and the rush of entrepreneurs to supply the demands of farmers for these fertilizers all attest to the sudden popularity enjoyed by commercial fertilizers.

Fertilizer Use Moves West

The use of commercial fertilizers began to spread westward from the Atlantic seaboard during the 1870s. In 1874 a farmer in Cayuga County in western New York reported that there were very few farmers in his area who did not use superphosphate on their wheat.[48] Another report from the area mentioned that the first experiments with superphosphates had begun only eight years earlier, but that the product was so markedly effective that its use would probably double or triple in the next few years.[49] In 1883 Joseph Harris reported that the use of superphosphates was greatly increasing in western New York.[50] Farther west, the lack of fertilizer usage contrasted sharply with the situation along the eastern seaboard. In the Midwest the 1880 census found only a few scattered pockets of moderate use of purchased fertilizers, generally near urban centers. In 1874 a Fairfield County, Ohio, farmer reported in the *American Agriculturist* that superphosphate had not been seen in his area.[51] In 1875 Joseph Harris told Wisconsin farmers that corn was much too cheap there and manures "too dear" for any artificial manure to be used successfully.[52] In 1876 a professor at the Kansas State Agricultural College in Manhattan, Kansas, indicated that even stable manure was "held in very light estimation" in that area. Instead of being used on fields, it was simply dumped in ravines or empty lots.[53] Nevertheless, as the soils of these areas became depleted during subsequent decades, they too would follow the lead of the Northeast in the use of commercial fertilizers.

Contrasting Patterns in the Southeast

The pattern of fertilizer usage in the Southeast was very different from that of the North. According to the 1880 census, farmers in the older cotton-producing areas of the Carolinas, Georgia, and Alabama consumed over one-third of the fertilizer used in the nation. Unlike in the North, the areas of heaviest usage were not concentrated around cities, but instead were spread in the interior. Census reports exaggerated the extent of use in the Southeast since the cash price in the South was 20 percent higher than in the North and the credit and cotton option systems added another 20 to 30 percent to the cost of fertilizers for southern farmers.[54] Nevertheless, even after discounting the effects of higher prices on census statistics, the area was a heavy user of commercial fertilizers.

The fertilizer system in the South had not developed out of urban-rural recycling systems as it had in the North, nor did it develop in response to the growing demands of urban areas for bulky or perishable farm products. Instead, as the historian Rosser H. Taylor concluded, in the Southeast after the Civil War, farmers took up fertilizers out of desperation caused by the loss of slave labor, the exhaustion of the land, and the introduction of sharecropping.[55] The small number of stall-fed stock resulted in a scarcity of domestic manures which made it even more difficult to maintain soil fertility.[56] In addition, as Roger L. Ransom and Richard Sutch have noted recently, the effective decrease in the southern labor force caused by the choice of freedmen and their families to work less than they had under slavery reduced the amount of land that could be worked by as much as 20 percent.[57] Thus, to maintain yields it was necessary to increase productivity of the land still under cultivation. Given the depleted condition of most land in the older cotton areas and the lack of alternative ways to increase productivity, increasing the fertility of the soil by the purchase of commercial fertilizers was a logical alternative.

Commercial fertilizers were the easiest short-term solution to the region's productivity problem. Effective programs of soil building based on crop rotation and green manuring would have required considerable time and capital investment. The sharecropping system, which had spread rapidly through the cotton belt after the war, however, did not include any provision for security of tenure or long-term soil improvements. Therefore, farmers were not willing to make investments in fertility unless they could expect a significant return during the first year.[58]

Fortunately for the South, commercial fertilizers, which had been developing for decades in the North, became available at just the time when they were most needed. The combination of sharecropping and the prevalence of farming on credit further encouraged the rapid adoption of commercial fertilizers since they promised landowner, creditor, and sharecropper the largest immediate return on investments. Moreover, the creditors and landlords could provide the capital necessary for the purchase of commercial fertilizers, but were not able or willing to provide the much more extensive capital and labor necessary for long-term soil-building programs.[59]

To some extent the effect of these fertilizers in the South was the opposite of that in the North. While northern farmers generally used commercial fertilizers to supplement domestic manures in a balanced rotation program designed to maintain or enhance the fertility of their farms, southern sharecroppers used fertilizers to help exploit their already run-down fields. Agricultural experts strongly recommended less reliance on commercial fertilizers and more emphasis on improved rotations, deep plowing, and domestic manures.[60] Southern farmers, however, continued to do exactly the opposite. This divergence of attitudes toward fertilizer use can be partly explained by differing economic circumstances in the two regions. The history of fertilizer use in each section, however, explains much of the difference. Northern farmers began using commercial fertilizers as a supplement to their recycling system, designed to facilitate long-term soil improvement.[61] In contrast, historically, southern farmers did not have a farming system designed to save their soil. Instead, in their system soil exploitation was an accepted part of farming. With neither the economic incentive to build up soil fertility on a long-term basis nor a heritage of carefully recycling all local wastes toward that goal, it is not surprising that the large-scale consumption of commercial fertilizers in the South did nothing to restore the region's exhausted soils. It may actually have contributed to their further deterioration by allowing more intensive exploitation of the land than would have been possible without the use of fertilizers. In contrast, many northern agriculturalists figured that moderate expenditures for commercial fertilizers would start a cycle of improvement which would result in the increased production of domestic manures and the permanent improvement of a farm.[62]

After 1880

By 1880 the fertilizer industry had evolved to its modern shape. Manufacturing techniques had been developed that could produce a reliable, consistent, relatively affordable fertilizer containing the proper balance of ingredients. Manufacturers made further improvements in their processes during the following decades, but these mainly changed the efficiency of the operation, not its general outlines. Sufficient supplies of raw materials had been developed to supply the major nutrients needed in fertilizers. After 1887 Florida and Tennessee phosphate deposits supplanted South Carolina deposits, but did not produce any major changes in the shape of the industry.[63] In the 1890s steel manufacturers began selling "basic slag" as a phosphatic fertilizer, but production never reached a significant scale in the United States.[64] The industry continued to scavenge for its nitrogen supply, 90 percent of which in 1900 still came from organic wastes. Not until the early twentieth century were techniques developed for the production of ammonia as a by-product of coke production and for the atmospheric fixation of nitrogen.[65] Germany continued to monopolize the supply of potassium until World War I, when a wartime embargo cut off the supply, encouraging the development of domestic supplies.[66] Despite these shifts in the sources of some of the raw materials, the general shape of the industry has remained basically unchanged. It still relies on mineral sources of phosphorus and potassium. The only significant change in the twentieth century has been the gradual substitution of inorganic sources of nitrogen for organic sources, completing the decline of the urban-rural recycling system as a source of plant nutrients.

Fertilizer firms, like many other kinds of American businesses, went through a merger movement at the end of the nineteenth century. Some segments of the industry attempted to form a trust at least as early as 1888.[67] At the end of the century, William Bradley of Boston and John Gibbons of Baltimore led a merger of many of the major fertilizer companies in the northern half of the country to form the American Agricultural Chemical Company, which survives today as the Agrico division of the Williams Companies.[68] About the same time, the larger companies in the southern states merged into the Virginia-Carolina Chemical Company, which is now part of the Mobil Oil Corporation.[69] The Virginia-Carolina Company may have been part of an attempt to monopolize the fertilizer business in the South that the Justice Department broke up in a 1906 antitrust suit.[70] In 1908 J. P. Morgan attempted to form the $50 million "United States Agricultural Corporation," but, failing in this attempt, put together the International Agricultural Chemical Company which survives today as the International Minerals and Chemical Corporation.[71]

The merger movement did not, however, drastically alter the fertilizer industry's structure, since the relative ease of entry, the advantages of smaller-scale production facilities located nearer their markets, and the wide variety of resources that can be used in fertilizer manufacture have prevented any single firm or small group of firms from dominating the market.

Not only had the fertilizer industry reached its modern shape by the 1880s, but farmers, at least along the eastern seaboard, were utilizing and thinking about fertilizer in basically the same ways that agriculturalists do today. Despite some resistance and the continuation of older practices, those farmers already considered the purchase of fertilizer as an essential and normal part of their operation, and were using fertilizers as an integral part of their agronomic system.

This transition severely weakened the long-standing predeliction of many farmers to be as self-sufficient as possible. Maximum possible self-sufficiency did not imply lack of participation in the market system. Rather it implied a strong preference for producing as much as possible at home, instead of maximizing production of crops best suited to the area and purchasing everything else. Commercial fertilizers offered such a significant advantage to many farmers, however, that they were willing to forgo the self-sufficiency that reliance on the on-the-farm recycling system allowed, and to take advantage of the increased profits available from the purchase and use of commercial fertilizers. Many eastern farmers were behaving like their urban entrepreneurial counterparts. Their goal was maximization of profits. If this could be reached only by large expenditures on raw materials, then that was what they would do.

Because of the transformations in the fertilizer industry and on the farm, the urban-rural recycling system, which had been such an important catalyst in the modernization process, had largely disappeared. The concept of recycling did not disappear, and manufacturers continued to use recycled wastes where economically feasible, but recycling concerns were rapidly diminishing. Thus, farmers and fertilizer manufacturers, like most of twentieth-century society, gave up on recycling as a viable alternative to a technological system dependent on resource exploitation and energy consumption. Nevertheless, even though it did not survive, the urban-rural waste recycling system that had developed in the middle decades of the nineteenth century shaped the present system of use and manufacture of fertilizers, and the legacy of the recycling system can still be seen today in industry and agronomic practices that have changed remarkably little in the years since 1880.

CHAPTER X

A Complex
Technological System

The evolution of the urban-rural recycling system into the fertilizer industry followed patterns common to the development of many complex technological systems where changes generally result from attempts to maintain the stability of the existing system. Instabilities caused by problems with soil fertility often initiated changes, since farming practices and environmental factors constantly threatened to deplete the fertility of all but the richest soils. This instability prompted the creation of the recycling system at the close of the eighteenth century in areas of the eastern United States where many years of cultivation had exhausted the soil. The old exploitative system was stable whenever there were sufficient supplies of new land and labor to clear that land, but where access to new land was lacking and labor was scarce or expensive, farmers in otherwise viable agricultural areas had to rehabilitate their soil by replenishing the nutrients removed by crops. Generally, these farmers turned first to local supplies of organic fertilizers. Once they had made the commitment to maintain or rebuild their soil's fertility, they readily substituted other fertilizers in reaction to new instabilities which began to develop in the system until the present "mature" fertilizer system had developed.

Instabilities and the Process of Substitution

The fertilizer system evolved first near the urban centers of the Northeast, where the combination of soil exhaustion, land scarcity, and strong market demand was most pressing, and then gradually spread to other parts of the country. By the 1830s the exhausted farm regions of eastern Maryland and Virginia began to follow the lead of the Northeast. In the late 1850s farmers in large areas of the Southeast began to face the same constraints. After the Civil War, the peculiar economic relations of the tenancy system exacerbated the region's triple problem of declining fer-

162

tility, lack of new land, and scarcity of labor. In the last decades of the century, the soil of the Ohio Valley began to lose its fertility, forcing farmers there to use artificial means of restoring the productivity of their soil.

The efforts of the farmers in the Northeast who first confronted the problem led to the evolution of urban-rural recycling systems that subsequently gave birth to the commercial fertilizer industry. Regions where the use of fertilizers began later generally borrowed the technologies already developed in the Northeast. Often northern firms, capital, and expertise moved south or west to facilitate the change. Because of this pattern, the technologies that had developed in the urban-rural recycling systems around northeastern cities in the first half of the century had a major influence on the development of the fertilizer system in the entire country during the second half of the century.

Another instability that often beset the fertilizer system was the shortage and consequent high price of materials. Repeatedly, the inability to produce or obtain sufficient fertilizing materials at a reasonable price frustrated farmers. The shortages of yard manure in farms near East Coast cities prompted the initial forays into the expanded local recycling system and then the urban-rural recycling system. The even more severe shortages of domestic manure later in the Chesapeake region, and still later in the South, forced farmers in those areas to enter the commercial fertilizer system. Insufficient supplies of bones, Peruvian guano, and phosphatic guano each caused searches for substitutes. Not only the farmers, but also the entrepreneurs who supplied them, actively sought acceptable alternatives. These efforts led to discoveries of new raw materials, such as phosphatic guano and South Carolina rock phosphates, and to manufacturing innovations, such as the production of poudrette, artificial guano, fish guano, superphosphates, and animal fertilizers.

Once an embryonic system of fertilizer production, distribution, and use was established in any region, subsequent instabilities, often generated within the system but sometimes caused by external factors, encouraged the further evolution of the system. The use of nutritionally unbalanced fertilizers produced one often-repeated pattern of instability. Long Island farmers who made extensive use of ashes early in the century soon discovered that after a few applications their effect was much diminished. Pennsylvania farmers using gypsum made the same discovery, as did farmers in most regions who attempted to use lime without any other fertilizers. Even farmers in the border states who used Peruvian guano for long periods began to claim that it was exhausting their land. In most cases the "law of the minimum" first formulated by Liebig was at work. Long Island fields responded to the supply of potassium or calcium in the ashes until some other nutrient was depleted so that it became the limiting factor in crop production. The sulfur in gypsum probably produced a similar effect in Pennsylvania. In many areas lime initially made soil nutrients more available to plants, but as soon as these nutrients were used up, subsequent applications produced no noticeable effect, and the soil appeared worse than at first. Similarly, the high ratio of nitrogen to phosphorus and potassium in Peruvian guano sometimes exhausted available reserves of the latter two elements in the soil.

In all of these instances, an initially successful fertilizer ultimately failed, causing a search for substitutes. Often the symptom was soil exhaustion similar to that which had prompted farmers to undertake the use of commercial fertilizers in the first place. When these situations occurred early in the century, farmers were at a loss to explain them. By the middle of the century, however, farmers and agricultural experts were beginning to understand the importance of a nutritionally balanced fertilizer. Whether or not farmers understood the reasons for the failure of a previously successful technology, the instabilities such a failure caused in the system encouraged farmers, entrepreneurs, and agricultural experts to seek substitutes. This search often led to significant evolutionary changes in the system. Sometimes these changes were in agronomic practices, such as not using lime alone. Sometimes these changes were substitutions of new fertilizer ingredients, such as the replacement of ash by bone. Sometimes these changes were the development of new types of fertilizers or of new processing technologies, such as the development of manipulated guanos and fish fertilizers in the late 1850s.

Other factors also created instabilities within the system. Some were endogenous to the system, such as inappropriate application methods used by farmers, prevalence of fraud and adulteration, and undetected variations in raw materials. Other instabilities were caused by exogenous factors such as the Civil War. Variations in climate and soil were another continual source of instability. Experiments that worked on one field one year were not successful another year or on another field. The resulting uncertainty delayed the adoption of fertilizers by some farmers. Whatever the source of the instability, the subsequent attempts to restore stability often resulted in significant, although unintended, alterations to the system of fertilizer manufacture, distribution, and use.

Generally, disruptions were met by attempts to substitute new fertilizing materials as similar as possible to those already in use, and requiring the minimum possible alterations in channels of supply, processes of production, methods of use, and mentalities of the farmers. Although the strong predilection to find similar replacements strongly influenced the choice of possible substitutes, economic considerations, energy requirements, and information networks were also important. Once a substitution had been made, it often produced new instabilities that necessitated still further evolutionary alterations to the system. For instance, the substitution of phosphatic guano for Peruvian guano forced dealers to build plants to manipulate or acidulate the phosphatic guano. By this process an on-the-farm closed-cycle system in the beginning of the century evolved through the urban-rural recycling system at mid-century to the energy- and resource-intensive, nonrecycling commercial fertilizer system in place by the end of the century.

Even linguistic evidence attests to the process of substitution involved in the evolution of the modern fertilizer system. Early in the nineteenth century, as farmers substituted other wastes for animal manure, they persisted in using the term "manure" to describe diverse fertilizers such as city street sweepings, ground bone, swamp muck, and waste ashes. Similarly, after the successful introduction of "poudrette" (a dry fertilizer made from human excrement) in the 1840s, promoters of Peruvian guano sometimes referred to their product as "natural poudrette," and manufacturers referred to newer products as "poudrette" despite substitutions of slaughterhouse refuse and other urban wastes as ingredients in fertilizer manufacture. The most striking linguistic evidence of the process of substitution was the persistence of the word "guano." Once farmers appreciated the value of Peruvian guano in the 1850s, entrepreneurs introducing substitutes generally called their article "guano"— "artificial guano," "fish guano," or "animal guano." Many companies continued to call their fertilizer "guano" during the last decades of the nineteenth century even though these mixtures contained no real guano. Farmers, especially in the South where the use of guano had the largest impact, used the word "guano" as a synonym for fertilizer well into the present century.

The Evolutionary Nature of the System

The evolutionary process that produced the modern fertilizer system late in the nineteenth century was the product of repeated substitutions that attempted to maintain the stability or improve the performance of the existing system. Early in the century farmers substituted locally obtainable materials such as leaves, fish, muck, or marl for scarce supplies of barnyard manure. Later they substituted purchased supplies of ashes, bones, street sweepings, stable manure, and other urban wastes. Then some farmers substituted Peruvian guano for these materials. Entrepreneurs substituted phosphatic guano for Peruvian guano, South Carolina phosphates for phosphatic guano, and German kainite for unleached wood ashes. They also developed, manufactured, or processed fertilizers such as superphosphates, poudrette, fish guano, and animal guano as substitutes for previously used fertilizers or as ways to make effective fertilizers out of substitute raw materials. Many similar substitutions led to the evolution of the modern fertilizer industry and practices.

Plotting all of these substitutions on a time line produces an evolutionary chart as in figure 10-1. Unlike biological evolution, new fertilizers were not necessarily direct descendants of previous ones. Sometimes this was the case, as was the development of the superphosphates industry from the use of untreated bones, but more often new developments were actually substitutions for earlier materials. Many innovations such as the fish guanos or superphosphates grew out of more than one antecedent. Although the chart does not adequately show the rate at which these developments occurred, as in biological evolution, many of the evolutionary changes in the fertilizer system were rapid, followed by long periods of stasis.

A process very much like natural selection was at work in the evolution of the fertilizer system. Although this diagram does not adequately demonstrate the process, the demands of growing plants for the principal fertilizing nutrients repeatedly led to the development of balanced fertilizers. Barnyard manure, street sweepings, poudrette, and the many varieties of composts were all complete fertilizers, although the concentrations of nutrients were low. Peruvian guano and ground bones were also relatively complete manures, although the balance between nutrients was not ideal. After the introduction of artificial fertilizers, biological factors forced manufacturers to offer balanced fertilizers and encouraged farmers to use combinations of fertilizers to achieve the same effect.

Despite these shortcomings, this chart clearly shows the large number of alternative lines of evolutionary development. Many of the paths were blind alleys. Some "archaic" forms persisted for long periods, side by side with newer developing technologies, for economic or cultural reasons. Some relatively insignificant fertilizers such as the phosphatic guanos led to highly significant new branches of evolutionary development. This chart also demonstrates that inducements for technological changes tended to produce simultaneous innovations in a number of different branches of the evolutionary tree. Although this chart relates only to the fertilizer system, similar diagrams could be drawn for other evolving technological systems.

The System's Information Network

The flow of information played an important role in the evolution of the system of fertilizer supply and use. Since technology is often defined as information, the network in which the information travels helps control the development of a technological system. An information loop paralleled most legs of the nutrient cycle. Some information traveled through the system in the opposite direction from the nutrient flow, such as information about the demand of urban areas for farm products (containing nutrients and energy) and the demand of farmers for fertilizers. Much of this information took the shape of money and can be conveniently measured in the prices of various commodities and fertilizers. It was this counterdirectional information that reinforced the system and led to its elaboration. Urban demands for agricultural products provided some of the initial demands from farmers for commercial fertilizers. These demands, in turn, encouraged entrepreneurs to develop the capacity to produce or obtain the needed products.

Perhaps the most important flow of information was from the plants in the farmer's field to the farmer who watched them grow. The amount and quality of a growing crop conveyed a great deal of information to farmers about the efficacy of the fertilizers they had applied. Farmers perceived destabilizing factors such as the reduced efficacy of ashes or gypsum through this channel. This was, however, a very imperfect channel for the transmission of information, especially in the early parts of the century. Farmers perceived only the increase or decrease in crop yields following the application of a fertilizer. Even these perceptions were often inaccurate, since they generally were not based on careful measurement of crop yields; and extraneous factors such as variations in soil, climate, disease, and plant genetics distorted trial results.

Nevertheless, the information the farmer received through this imperfect channel was highly influential in the decisions he made about his subsequent efforts to restore soil fertility. He was inclined to repeat whatever worked without too many questions. Whenever failures occurred or the need for substitutions arose, the recycling mentality and not scientific knowledge generally determined what changes would be made, especially in the early part of the century. Since many of the substitions suggested by the recycling mentality were effective, that mentality was generally reinforced.

These two information channels associated with the nutrient cycle were part of a much larger information system in which the principal information flows were among the farmers, agricultural journals, dealers, manufacturers, scientists, and agricultural experts. Farmers, and to a lesser extent dealers and manufacturers, linked this part of the information system with the information channels along the nutrient cycle. In addition to talking with and observing their neighbors, farmers communicated through the journals and agricultural societies and clubs.

The scientists and experts received information from farmers, processed it, and in return provided scientific information that often reached farmers through the journals. With this new information, farmers more accurately perceived what was happening when they applied fertilizers to their soil. This made the substitutions chemically more precise, although late in the nineteenth century this process was still very imperfect. Since soil chemistry and microbiology are exceedingly complex subjects, only poorly understood today, early in the nineteenth century it was the

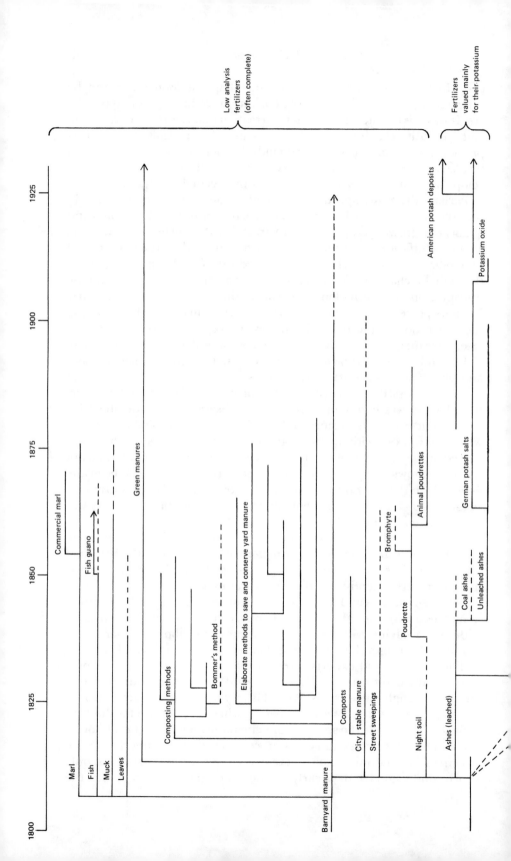

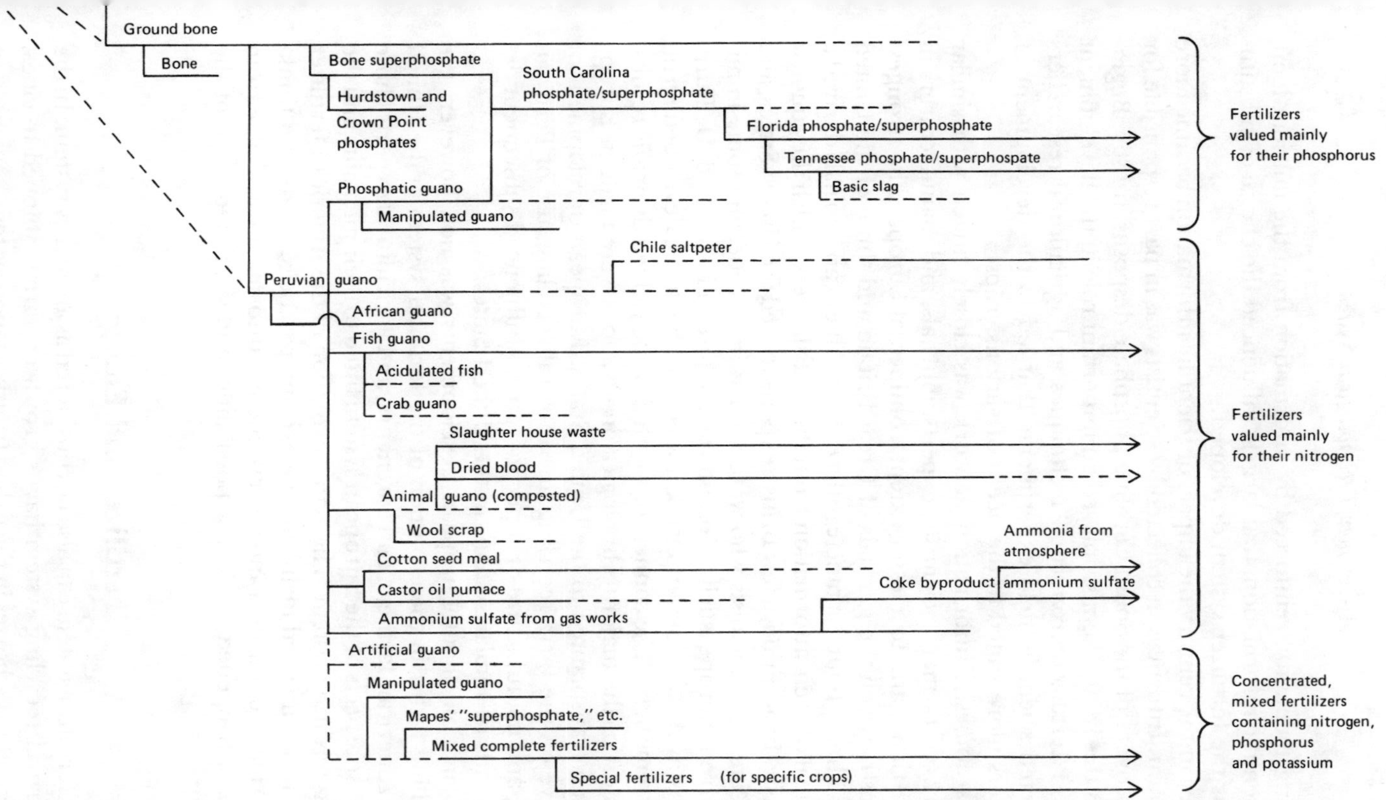

Ground bone
Bone
Bone superphosphate
Hurdstown and
Crown Point
phosphates
South Carolina
phosphate/superphosphate
Florida phosphate/superphosphate
Tennessee phosphate/superphospate
Basic slag
Phosphatic guano
Manipulated guano
Chile saltpeter
Peruvian guano
African guano
Fish guano
Acidulated fish
Crab guano
Slaughter house waste
Dried blood
Animal guano (composted)
Wool scrap
Cotton seed meal
Castor oil pumace
Ammonium sulfate from gas works
Coke byproduct ammonium sulfate
Ammonia from
atmosphere
Artificial guano
Manipulated guano
Mapes' "superphosphate," etc.
Mixed complete fertilizers
Special fertilizers (for specific crops)
Fertilizers
valued mainly
for their phosphorus
Fertilizers
valued mainly
for their nitrogen
Concentrated,
mixed fertilizers
containing nitrogen,
phosphorus
and potassium

Fig. 10-1. Diagram of the evolutionary development of the fertilizer system. The actual development was more complex than can be illustrated in this diagram. Many fertilizers had more than one antecedent. For instance, fish guano was a substitute for Peruvian guano, but it also developed from the earlier use of unprocessed fish as fertilizer. Similarly, Peruvian guano was a substitute for not only ground bone, but also barnyard manure and poudrette. Another relatonship not apparent in the diagram is that the mixed fertilizers were generally composed of various combinations of the potassium, phosphorus, and nitrogen fertilizers listed above them.

recycling mentality, reinforced by information from the nutrient loop, not scientific information, that largely influenced the choice of substitutions as the fertilizer system developed.

Later in the century the impact of scientific information became more important. Information collected by scientists was in part responsible for the discovery of the South Carolina phosphate desposits in 1867. Beginning in the 1870s chemists played an important role in the development of better fertilizer production techniques and agricultural researchers conducted scientific field experiments that led to the development of fertilizers more suited to the needs of various crops.

This domestic information network was closely linked with similar networks of farmers, scientists, experts, journals, and manufacturers in Great Britain, and to a lesser extent, continental Europe. The strongest link was between the journals of Great Britain and those of the United States. Throughout the nineteenth century, American journals derived a great deal of their information from the British press, which they quoted copiously. In addition, a few British experts toured the United States, and a few American chemists who were trained in Europe, and some immigrants from Europe (such as Joseph Harris, Joseph Lister, C. B. De Burg, and Gustav A. Liebig), supplied knowledge of the more developed fertilizer system there. The connections with Europe yielded dramatic results, since most of the major technological innovations were made in Europe and then transplanted to the United States. European development of intensive on-the-farm recycling, urban-rural recycling, use of Peruvian guano, and manufacture of superphosphates influenced subsequent developments of similar systems in the United States.

Information about European developments was not, however, the controlling factor in the evolution of the fertilizer system in the United States. American farmers, entrepreneurs, and agricultural experts chose only those aspects of the European system fitted to their mentality and the existing fertilizer system. The actual evolution of the American fertilizer system was controlled mainly by the existing practices, needs, and thinking of American farmers and entrepreneurs. Information available from Europe only opened a few new possibilities and suggested a few of the substitutions.

Fertilizer and Energy

The fertilizer system can also be considered as an energy system. In the urban-rural recycling system hay is grown on farms, shipped to cities, consumed by the horses in the urban transportation system, converted to manure, and shipped back to the farms. Only the nutrients were recycled, but the main purpose of recycling these nutrients was to help convert more solar energy into a form usable by the urban transportation system.

Without the nutrients, the grass and clover plants could not store enough solar energy in hay to power the city's horses. Plants converted solar energy into hay, but considerable energy was consumed and lost in the process. Harvesting, processing, and transporting the hay to the city consumed further energy. The horses of the city's transportation system consumed most of the remaining energy. The farmer's team, which carted the hay to the city and the manure back, consumed additional energy in the form of hay. A portion of the solar energy originally stored in the hay remained in the manure, but this was consumed by microorganisms in the soil or compost heap as they converted the nutrients into forms directly usable by growing plants.

The crucial difference between agricultural systems using fertilizers and the exploitative "soil mining" system of agriculture was the commitment of energy to recycle nutrients. On-the-farm recycling required both human and animal energy to cart manure from the farmyard to nearby fields. Indeed, this energy requirement was the principal reason why hard-pressed frontier farmers delayed using their manure for as long as possible. (Certainly it was not lack of knowledge of the utility of manure as a fertilizer.) The successive enlargements of the system generally entailed the commitment of larger amounts of energy to supply plants with nutrients. Fishing on Long Island required much human and animal energy, as did digging and carting muck or marl. Transporting manure from cities also represented a large commitment of energy to the fertilizer system.

When hay was the principal form in which energy could be consumed, there were narrow limits to the extent the system could be enlarged. Hauling marl, muck, or stable manure more than five or ten miles probably consumed more energy than the resulting increases in crop production could justify. One solution to this problem was the substitution of concentrated fertilizers, which required less energy to transport. The more important innovation, however, was the substitution of less expensive forms of energy, such as wind or fossil fuels, for the hay that powered the early on-the-farm and urban-rural systems. The use of sailing ships and steam railroads not only made possible the efficient use of larger qualtities of fertilizers, it also allowed the urban-rural system to expand considerably beyond the five-or-ten-mile limit imposed by animal transportation of manure, and it allowed the use of distant raw materials such as Peruvian guano or German kainite. In addition, the application of power to the drying, grinding, acidulating, and mixing of fertilizer ingredients made possible the efficient production of concentrated fertilizers on a large scale.

As the system developed in the latter half of the century, although it used less human and animal energy, it required larger and larger energy supplements to keep it operating. While the application of concentrated fertilizers required considerably less energy on the part of the farmer, the mining and transporting of South Carolina phosphates and German kainite, the processing of fish guanos, the drying and grinding of slaughterhouse waste, the production of sulfuric acid, and the acidulation of bone or rock phosphates all required considerable energy—usually in the form of wood, coal, hay, water power, or wind. As a result of the gradual increase in the energy supplement to the nutrient cycle, the manufacture and use of fertilizer has developed into an energy-intensive system that has become a significant problem in the late twentieth-century world where energy has become scarce and expensive.

Connections with Cultural, Economic, and Political Systems

The fertilizer system and its various subsystems did not develop in isolation. It was closely connected with larger cultural, economic, and political systems. What I have called the "recycling mentality" was part of the national culture and linked to other aspects of it. Certainly the emphasis on avoiding waste was deeply rooted in the thinking of eighteenth-and nineteenth-century Americans, especially farmers. The recycling mentality was closely related to the virtues of frugality and self-reliance that developed of necessity in most rural areas. When the necessity of restoring the fertility of his soil became apparent to a farmer, he was much more comfortable with a closed-cycle nutrient system that mirrored his self-sufficient operation than he would have been with the commercial system that developed later, in which farmers sold a certain portion of their produce in the market in order to afford the large quantities of commercial fertilizers they needed.

Once developed, the recycling mentality influenced cultural perceptions. On the local level, farmers in northern areas where the system was best developed often judged the character of neighbors by their ability to make the on-the-farm system work. On a larger scale, a number of writers connected recycling with concepts of natural regeneration or dissipation. They argued against exportation of exhausting crops without the corresponding importation or domestic production of fertilizers to replace the nutrients sent abroad. For instance, Horace Greeley, the most widely read of the proponents of this theory, argued in 1870 that the continued exportation of wheat was gradually impoverishing the nation's soil and robbing its vitality.[1]

The development of the fertilizer system was intimately connected with urban growth. The rapidly growing East Coast cities were the principal foci of the nutrient cycles and informational networks described here. Nearby urban markets were prime incentives for the development of intensive farming (including the use of fertilizer). In return, the use of recycled urban wastes as fertilizer allowed the farmers to produce the fruit, vegetables, poultry products, and hay those cities needed, but which could not be transported great distances. Without this support, urban growth might have been checked or deflected at some point in the nineteenth century.

In addition, the development of the fertilizer system may have significantly affected migrational patterns. Without the introduction of commercial fertilizers, agriculture would have been doomed in many areas of the Northeast. Many more rural people might have attempted to migrate to the cities or the West than actually did. Moreover, those who remained on impoverished farms might have been as frustrated with the political system by the end of the century as were western and southern Populists, instead of being staunch supporters of the conservative eastern political establishment. The impact may have been even more dramatic in areas of Maryland and Virginia where the introduction of commercial fertilizers reversed the agricultural decline and turned some sections from sources of migration to the West into targets of migration from the North. It is difficult even to guess what might have happened demographically in the South after the Civil War if productivity had fallen much lower because of the lack of fertilizers to sustain agriculture on the area's marginal lands.

There may have been some connections between the development of th fertilizer system and sectional politics. The much more intensive development of the recycling system in the North before the Civil War, especially the on-the-farm portion of the system, led to widespread criticism of southern agricultural practices by northern farmers and agricultural experts, and may have contributed to the free-soil ideology before the war and to continuing antisouthern feelings after the war.

To some extent, the difficulties of developing such a system in the South may have been one of the factors which made the slave labor plantation system and its extensive methods a viable alternative to the much more intensive practices followed on most northern farms. As Avery Craven has noted, the introduction of commercial fertilizers into Virginia and Maryland in the 1840s and 1850s caused the rural economy of the area to become much more like that of the North and encouraged many emigrants from the North to settle in the two states.[2] If this process had continued for another decade or two before the outbreak of the Civil War, the primary attachment of the region might have been more toward the North than the South.

The introduction of guano, which played such a large role in the development of the fertilizer system, also had international repercussions. It gave the United States its first taste of the problems of dealing with a less developed nation with an organized monopoly on an essential raw material, resembling the oil cartel of the Organization of Petroleum Exporting Countries (OPEC) in the 1970s. Several administrations attempted to alleviate the problem by diplomatic means, but to no avail. When entrepreneurs tried to outflank the Peruvian monopoly by locating new sources of guano, they caused a new series of diplomatic disputes and gave the nation a preview of the problems of imperialism. Only the discovery of domestic phosphate supplies and the development of a fertilizer industry capable of manufacturing substitutes for guano from domestic resources ended the problems caused by the Peruvian monopoly. By the end of the century, however, the German potash cartel began to create similar problems.

On the state and local level, the development of the fertilizer industry led to demands in the second half of the century for greater government regulation of businesses. Residents offended by the obnoxious fumes of nearby fertilizer plants demanded and generally received relief from sympathetic local authorities. Farmers faced with the problems of massive fertilizer fraud and a generally unreliable product demanded and received laws requiring state inspection and regulations for the fertilizer industry. In addition, the need for testing and experimenting with fertilizers was one of the initial arguments for the first state agricultural experiment stations.

Finally, the legacy of the ultimate failure of recycling urban wastes to supply the nutrients needed by nineteenth-century agriculture can be seen in the attempts by today's environmentalists to restore precisely such a system. As the world's demand for food grows, as energy costs become larger, and as mineral resources are gradually depleted, many see a return to a recycling system as the only viable alternative. Certainly the nineteenth century made a serious effort to make such a system work. The urban wastes that were most successfully recycled in the nineteenth century, slaughterhouse wastes and fish, are now being processed into animal feed, a less energy- and resource-efficient use. Attempts to recycle other materials, expecially the garbage and sewage, fell victim to strong economic and public health considerations. Moreover, recycling bulky organic materials requires immense consumption of energy, as every nineteenth-century farmer who carted muck or manure knew. Attempts to turn back the clock face severe impediments—the very same problems that originally caused the substitution of energy- and resource-intensive commercial fertilizers for the recycled organic wastes previously used.

Appendix

Companies Producing Superphosphates Before 1869.*

Company	Location	Year
PROTO-SUPERPHOSPHATES		
P. S. Chappell	Baltimore	1849
Kettlewell & Davison	Baltimore	1849
George H. Barr	New York City	1851
SUPERPHOSPHATES		
J. J. Mapes	Newark	1852
C. B. De Burg	Brooklyn	1852
William Patterson	Newark	1853
Eagle Chemical Works	New York City	1853
C. B. Rogers (Rogers & Best)	Philadelphia	1854
New York Superphosphate		
Manufacturing	New York City	1854
Andrew Coe	Middletown, CT	1854
Baugh & Sons	Philadelphia	1855
New Orleans Bone Black Company	New Orleans	1856
B. M. Rhodes	Philadelphia and	
	Baltimore	1856
Hildreth	New York City	1856
Lawson S. Hoyt	New York City	1856
Potts & Klett	Philadelphia	1850s
Buck	Hartford, CT	1857
New Jersey Superphosphate	New Jersey	1857
Coe & Company	Boston	1857
L'loyd	Providence	1857
Allen & Needles	Philadelphia	1857
Hathaway	Boston	1857
Pike & Company	Connecticut	1857

Company	Location	Year
American Fertilizing Company	New York City	1857
William Trego	Baltimore	1857
William Whitlock	Baltimore	1858
Green & Preston	Connecticut	1858
Tasker & Clark	Philadelphia	1859
Joshua Horner	Baltimore	1859
Moro Phillips	Philadelphia	?
Lewis M. Hatch	Charleston, SC	1860
Fowle & Company	Alexandria, VA	1860
Richmond Fertilizer Manufacturing Mills (Hartman)	Richmond, VA	1860
E. Frank Coe	New York City	1860
Lister Brothers	Tarrytown, NY (later Newark)	1860
William Bradley	Boston	1861
Maryland Fertilizer and Manufacturing Company	Baltimore	1863†
Pacific Guano Company	Woods Hole, MA	1863†
Cumberland Bone Company	Boothbay, ME	1864†
Patapsco Guano Company	Baltimore	1865
George E. Currie	Cincinnati	1865
Armor Smith	Cincinnati	1865
(Unidentified)	Richmond	1865
T. T. Hubbard & Son	Charleston, MD	1866
Davison, Symington & Company	Baltimore	1866
G. F. Ober	Baltimore	1867
Runford Chemical Works	Providence	1867
Henry Bower	Philadelphia	1867
Chesapeake Guano Company	Baltimore	1868

*For sources, see notes 82, 83, 84, and 93 for Chapter VII.

†Dates are generally those at which a firm commenced manufacturing superphosphates. Dates for the Pacific Guano Company, Maryland Fertilizer and Manufacturing Company, and Cumberland Bone Company are organization dates. Production probably began a year or two later.

Notes

In citing works in the notes, short titles have generally been used. The following abbreviations have been used:

AMOGMM   Association of the Menhaden Oil and Guano
                Manufacturers of Maine
MSBA   Massachusetts State Board of Agriculture
NYSAS   New York State Agricultural Society
NYSPUA   New York Society for the Promotion of Useful
                Arts

Chapter I: The Recycling System

1. For example, see Duc de La Rochefoucauld-Liancourt, *Travels through the United States of North America*, 2nd ed., 4 vols. (London: R. Phillips, 1800), 2: 275, 309; 3: 24, 146–47, 484, 511; 4: 125; Richard Parkinson, *A Tour of America in 1798, 1799, and 1800* (London: J. Harding, 1805), 197–201; William Strickland, *Journal of a Tour in the United States of America, 1794–1795*, reprint ed. (New York: New York Historical Society, 1971), 99, 133, 176–78; Robert Barclay-Allardice, *Agricultural Tour in the United States and Upper Canada* (Edinburgh: William Blackwood & Sons, 1842), 137–38.

2. For discussion of the nonrecycling, exploitative system, see Percy Wells Bidwell and John I. Falconer, *History of Agriculture in the Northern United States, 1620–1860* (Washington: Carnegie Institution of Washington, 1925), 87; Paul W. Gates, *The Farmer's Age: Agriculture, 1815–1860* (New York: Holt, Rinehart and Winston, 1960), 102; Neil Adams McNall, *An Agricultural History of the Genesee Valley, 1790–1860* (Philadelphia: University of Pennsylvania Press, 1952), 113–14; Hubert G. Schmidt, *Agriculture in New Jersey: A Three-hundred Year History* (New Brunswick, N.J.: Rutgers University Press, 1973), 59–79.

3. Culhbert W. Johnson, *On Fertilizers* (London: Ridgway, Piccadilly, 1839), 1–30, was well known in the United States. *Cultivator* 3 (1836–37): 3–4, 23; 7

(1840–41): 86; Gouverneur Emerson, *Address Delivered before the Agricultural Society of Kent County, Delaware* (Philadelphia: National Merchant Printer, 1857), 13.

4. *Genesee Farmer,* quoted in *American Farmer* 14 (1832–33): 339; *American Farmer* 12 (1830–31): 275; *Cultivator* 1 (1834–35): 29; 4 (1837–38): 43; n.s., 7 (1850) 386–87.

5. Benjamin D. Cooper, letter in *American Farmer* 9 (1827–28): 411.

6. *Cultivator* 5 (1838–39): 144.

7. Johnson, *On Fertilizers,* 1–30; *Cultivator* 3 (1836–37): 3–4, 23; 7 (1840–41): 86; Emerson, *Address,* 13.

8. *American Farmer* 3 (1821–22): 169.

9. John Claudius Loudon, *An Encyclopaedia of Agriculture* (London: Longman, Hurst, Rees, Orne, Brown, and Green, 1825), quoted in *American Farmer* 9 (1827–28): 9.

10. *American Farmer* s. 3, 2 (1840–41): 165.

11. *Cultivator* 2 (1835–36): 1.

12. Charles T. Jackson in U.S. Patent Office, *Agricultural Report* (1844), 378.

13. John A. Dix, address, NYSAS, *Transactions* 19 (1859): 31.

14. Frederick L. Olmstead, *A Journey in the Seaboard Slave States* (New York: Mason Brothers, 1856), 168; *Southern Planter* 15 (1855): 163.

15. For examples of the persistence of the recycling concept, see *Cultivator* s. 3, 8 (1860): 213; Emerson, *Address,* 13–17; *Plough, Loom, and Anvil* 10 (1857–58): 9; William Ford, *The Industrial Interests of Newark, New Jersey* (New York: Van Arsdale, 1874), 213–14.

16. *The American Farmer* n.s., 2 (1835–36) had thirteen articles on managing manure; s. 3, 1 (1839–40) had seventeen articles; s. 4, 3 (1847–48) had thirty articles; and the *Cultivator* n.s., 6 (1849) had twenty-nine articles. For examples see *American Farmer* 10 (1828–29): 217–19; 14 (1832–33): 211, 226; *Southern Planter* 1 (1841): 68; *Cultivator* 2 (1835–36): 12–14.

17. John Hannam, *The Economy of Waste Manures: A Treatise on the Nature and Use of Neglected Fertilizers* (London: Longman, Brown, Green, and Longmans, 1844): 32.

18. For example, see *American Farmer* 12 (1830–31): 300, 322, 361, 367; *American Agriculturist* 3 (1844): 39.

19. *American Farmer* 8 (1826–27): 122; 13 (1831–32): 185; *Cultivator* 7 (1840–41): 135, 152; Samuel L. Dana, *A Muck Manual for Farmers* (Lowell, Mass.: Daniel Bixby, 1842), 177–78.

20. Ezra L'Hommedieu, "Observations on Manures," NYSPUA, *Transactions,* 2nd ed., 1 (1801): 233; Hannam, *Waste Manures,* 17–64.

21. *Working Farmer* 3 (1851): 148.

22. *American Farmer* s. 3, 3 (1841): 381.

23. See George Bommer, *New Method Which Teaches How to Make Vegetable Manure,* 2nd ed. (New York: Redfield and Savage, 1845), 230; *American Agriculturalist* 3 (1844): 9; advertisement, *American Farmer* s. 4, 1 (1845–46): 319; *Cultivator* n.s., 1 (1844): 31.

24. *Cultivator* n.s., 2 (1845): 157; *American Farmer* s. 3, 3 (1841): 369; 5 (1823–24): 241; Richard A. Wines, "Agricultural Transition in an Eastern Long Island Community," *Agricultural History* 54 (1981): 54. The use of fish may have had its

roots in European practice instead of American custom. See Lynn Ceci, "Fish Fertilizer, a Native North American Practice," *Science* 188 (April 1975): 24–30.

25. *American Farmer* 1 (1819–29): 23; 13 (1831–32): 185, 201; *American Agriculturist* 1 (1842): 337; Ezra L'Hommedieu, "Experiments Made by Manuring Land with Sea Weed," NYSPUA, *Transactions* 1, pt. 2 (1794): 99; La Rochefoucauld, *Travels*, 2: 321.

26. For a brief discussion of the breaking of the "closed circuit system" in Great Britain, see F. M. L. Thompson, "The Second Agricultural Revolution, 1815–1880," *Economic History Review* s. 2, 21 (1968): 64.

27. *American Farmer* n.s., 5 (1838–39): 14.

28. Timothy Dwight, *Travels in New England and New York*, 4 vols. (1821–22; reprint ed., Cambridge: Harvard University Press, 1969), 3: 227. For other early descriptions of Long Island agriculture, see Strickland, *Journal of a Tour*, 41; John Harriott, *Struggles through Life*, 3rd ed., 3 vols. (London: Longman, Hurst, Rees, Orne and Brown, 1815), 184–85; H. G. Spafford, *A Gazetteer of the State of New York* (Albany: H. C. Southwick, 1813), passim.; Ezra L'Hommedieu, "Communications Made to the Society, Relative to Manures," NYSPUA, *Transactions* 1, pt. 1 (1792): 63.

29. Harriott, *Struggles through Life*, 187; La Rochefoucauld, *Travels*, 4: 243.

30. L'Hommedieu, "Communications," 57.

31. Dwight, *Travels*, 3: 213.

32. Ezra L'Hommedieu, "Experiments Made by Manuring Land with Seaweed Taken Directly from Creeks and with Shells," NYSPUA, *Transactions* 1, pt. 2 (1794): 99–102; NYSAS, *Transactions* 3 (1843): 464; *Cultivator* n.s., 5 (1848): 187; n.s., 6 (1849): 30–31.

33. *Cultivator* n.s., 5 (1848): 345. For an expanded discussion of this, see Wines, "Agricultural Transition," 54. See also, L'Hommedieu, "Observations on Manures," 231; *Cultivator* 6 (1839–40): 22.

34. Sometimes these were "still dairies" where the cows were fed the waste of neighboring distilleries. This may have been an efficient recycling system, but the milk was tainted and the establishments were especially foul. See *American Agriculturist* 2 (1843): 195, for a good description of one of these in Brooklyn.

35. For material on the sale of stable manure in New York, see NYSAS, *Transactions* 2 (1842): 207; 3 (1843): 462; L'Hommedieu, "Observations on Manures," 238; *American Farmer* 8 (1826–27): 340; John Duffy, *A History of Public Health in New York City, 1625–1866* (New York: Russell Sage Foundation, 1968), 63, 127, 185. In spite of the importance of the function, the men involved generally remained nameless. Dealing in manure and compost was apparently not considered respectable enough to be included in city directories.

36. For material on the removal of street dirt in New York City, see NYSAS, *Transactions* 2 (1842):207–208; Duffy, *Public Health, 1625–1866*, 176–94, 356–90. In addition, New York City, Common Council, *Minutes of the Common Council, 1784–1831*, reprint ed., 19 vols. (New York: City of New York, 1917), contains numerous references to street cleaning contractors, manure inspectors, etc.

37. According to Dwight, *Travels*, 3: 212, Long Island supplied a considerable portion of the fuel wood used in New York City.

38. *Long Island City Star*, quoted in *Sag Harbor Corrector*, 17 November 1866.

Also see Roberta Pincus, *This Is Great Neck, A History of the Great Neck Community from 1600 to the Present* (Great Neck, N.Y.: League of Women Voters, 1975), 25; and Udall family, Account Book, 1845–50, Nassau County Museum, East Meadow, N.Y., for numerous references for the manure trade to Great Neck. Udall purchased manure in New York, carried it on his sloop, and acted as a dealer for at least eleven of his neighbors. See *American Agriculturist* 19 (1860): 229, for statistics on manure carted by the Long Island Railroad. The following year, according to the *American Agriculturist* 20 (1861): 235, two million carman loads annually were carted from New York to Long Island.

39. NYSAS, *Transactions* 8 (1848): 530–32.

40. Ibid., 527–29.

41. *American Agriculturist* 20 (1861): 235.

42. NYSAS, *Transactions* 3 (1843): 464.

43. For a modern analysis of stable manures, see Nyle C. Brady, *The Nature and Properties of Soil*, 8th ed. (New York: Macmillan, 1974), 538, 540. For typical rates of application, see NYSAS, *Transactions* 7 (1847): 632–36; 8 (1848): 529.

44. Chancellor Livingston, "Experiments and Observations on Calcarious and Gypsious Earths," NYSPUA, *Transactions* 1, pt. 1 (1792): 51; L'Hommedieu, "Communications," 71; *American Farmer* 9 (1827–28): 410; 14 (1832–33): 99.

45. NYSAS, *Transactions*, (1843): 465–66; 6 (1846): 634.

46. Livingston, "Experiments," 51; *American Farmer* 15 (1833–34): 76.

47. NYSAS, *Transactions* 6 (1846): 634; *American Farmer* 14 (1832–33): 99; *Cultivator* s. 3, 5 (1857): 142.

48. R. M. W., letter in *Genesee Farmer*, quoted in *American Farmer* 15 (1833–34): 93, mentioned that he had observed a Long Island farmer bring a load of cordwood to a landing, stack it, exchange it for a load of ashes, pay four dollars, and consider himself ahead!

49. NYSAS, *Transactions* 3 (1843): 465–66; *Cultivator* s. 3, 5 (1857): 142.

50. Leached ashes contain approximately 1.5 percent P_2O_5, 1 percent K_2O, and 28–30 percent lime (Gilbeart H. Collings, *Commercial Fertilizers: Their Sources and Use* [Philadelphia: P. Blakiston's Son, 1938], 234).

51. Ibid.

52. Brady, *Nature and Property of Soils*, 474.

53. *American Farmer* 15 (1833–34): 163; NYSAS 3 (1843).

54. The price of manure had doubled in twenty years, according to *Cultivator* 6 (1839–40): 121.

55. See Lodi Manufacturing Company, *An Act to Incorporate the Lodi Manufacturing Company . . . with Accompanying Remarks and Documents* (New York: H. Cassidy, 1840), 3–4.

56. *Cultivator* s. 3, 1 (1853): 226; *American Agriculturist* 7 (1848): 220.

57. The most important of these were J. J. Mapes's plant in Newark, New Jersey, and C. B. De Burg's plant in Williamsburg, Long Island.

58. *Long Island City Star*, quoted in *Sag Harbor Corrector*, 17 November 1866.

59. U.S. Census Office, Tenth Census [1880], vol. 3, *Report on the Productions of Agriculture* (Washington: GPO, 1883), 104–40.

60. NYSAS, *Transactions* 12 (1852): 636; "Agricola," NYSPUA, *Transactions*, 2nd ed., 1 (1801): 258–60; NYSAS, *Transactions* 2 (1842): 636.

61. See Gates, *Farmer's Age, 28; American Agriculturist* 2 (1843): 45; Strickland, *Journal,* 216–17.

62. Stevenson Whitcomb Fletcher, *Pennsylvania Agriculture and Country Life, 1640–1840* (Harrisburg: Pennsylvania Historical and Museum Commission, 1950), 124; La Rochefoucauld, *Travels,* 3: 484.

63. *American Farmer* 9 (1827–28): 297.

64. Fletcher, *Pennsylvania Agriculture, 1640–1840,* 128.

65. Richard Peters, "On Gypsum," Pennsylvania Society for Promoting Agriculture, *Memoirs* 1 (1808): 158; Richard Peters, *Agricultural Inquiries on Plaster of Paris* (Philadelphia: Cist and Mark, 1797), 68.

66. Fletcher, *Pennsylvania Agriculture, 1640–1840,* 137.

67. Peters, *Agricultural Inquires,* 10, 14, 17, 38, 43, 52.

68. Gerald S. Graham, "The Gypsum Trade of the Maritime Provinces: Its Relation to American Diplomacy and Agriculture in the Early Nineteenth Century," *Agricultural History* 12 (1938): 209–10.

69. Fletcher, *Pennsylvania Agriculture, 1640–1840,* 127, 136.

70. Ibid., 135.

71. Ibid., 136.

72. *American Farmer* 10 (1828–29): 345–46.

73. According to private communication from Dale E. Baker, Professor of Soil Chemistry at Pennsylvania State University, $CaSO_4$ (gypsum) may have produced such dramatic results because it helped maintain a more ideal Ca:Mg ratio in the soil over the short term. Eventually, however, phosphorus or nitrogen became a limiting factor and acidity increased to levels associated with aluminum toxicity. Professor Baker also speculates that the sulfur in gypsum may have been important in the period before the introduction of anthracite coal as an industrial fuel. Atmospheric discharge of sulfur by coal burning now supplies enough of that essential nutrient in the Northeast that artificial supplies are unnecessary.

74. Fletcher, *Pennsylvania Agriculture, 1640–1840,* 132.

75. *American Farmer* 12 (1830–31): 275. The "Hagerston" soil type, found extensively in Lancaster County, was naturally high in phosphorus, contained reasonable quantities of potassium, and was not reduced by leaching (USDA, *Atlas of American Agriculture* [Washington: GPO, 1935], 32–34, plate 5–8).

76. Henry Francis James, *The Agricultural Industry of Southeastern Pennsylvania* (Philadelphia, 1928), 22.

77. *American Farmer* 9 (1827–28): 403; 8 (1827): 233.

78. U.S. Patent Office, *Agricultural Report* (1849): 301.

79. According to Solon Robinson, *Guano: A Treatise of Practical Information for Farmers* (New York, 1852), 33, guano was less used in Pennsylvania because of the great improvements from the use of lime.

80. Pennsylvania State Agricultural Society, *Transactions* 1 (1854): 180.

81. For 1880 statistics, see U.S. Census Office, *Statistics of Agriculture, 1880,* 131; and map in figure 9–1.

82. For material on New Jersey, see *Cultivator* n.s., 1 (1852): 269–70; *American Farmer* 11 (1829–30): 293–95; 9 (1827–28): 411.

83. *Cultivator* 3 (1836–37): 21, 52; 8 (1841–42): 120, 182; *American Farmer* 15 (1833–34): 212; *American Agriculturist* 1 (1842): 374; 10 (1851): 124; Samuel W.

Johnson, *Essays on Peat, Muck, and Commercial Manures* (Hartford: Brown & Gross, 1859), 163–64; James F. W. Johnston, *Notes on North America: Agricultural, Economical, and Social* (Edinburgh: Blackwood, 1851), 306–309. See also Schmidt, *Agriculture in New Jersey*, 125; and Bidwell and Falconer, *History of Agriculture*, 234.

84. For statistics on the rate of application, see Henry D. Rogers, *Report on the Geological Survey of the State of New Jersey* (1836), quoted in D. J. Browne, *The Field Book of Manures or the American Muck Book* (New York: A. O. Moore, 1853), 121–29. For a modern analysis of greensand marl, see Collings, *Commercial Fertilizers*, 244. In the late 1850s, a group of entrepreneurs from Monmouth County and New York City attempted to exploit greensand marl commercially. They formed the New Jersey Fertilizer Company, but the effort was not successful, probably because the concentration of minerals in the marl was not high enough to pay the cost of transport (Johnson, *Essays*, 163–64; advertisement, *American Agriculturist* 18 [1859]: 127).

85. See Robinson, *Guano*, 37. For the 1880 pattern of use, see figure 9–1.

86. See Avery O. Craven, *Soil Exhaustion as a Factor in the Agricultural History of Virginia and Maryland, 1606–1860* (Urbana: University of Illinois, 1926), 129–31, for an analysis of the importance of transportation improvements in the area.

87. *American Farmer* 1 (1819–20): 98–99.

88. See Parkinson, *Tour*, 197–204. The best historical accounts of the problems of the area are: Craven, *Soil Exhaustion*, 128–63; Vivian Wiser, "Improving Maryland's Agriculture, 1840–1860," *Maryland Historical Magazine* 62 (1962): 105–32; Gates, *Farmer's Age*, 4–5, 107.

89. *American Farmer* 1 (1819–20): 98–99.

90. See Edmund Ruffin, *An Essay on Calcareous Manures* (1832; reprinted, Cambridge: Harvard University Press, 1961), 29.

91. Parkinson, *Tour*, 204. For similar arguments, see *Southern Planter* 15 (1855): 163–65.

92. *American Farmer* 13 (1831–32): 249.

93. Ibid. 13 (1831–32): 249; 11 (1829–30): 9–10; 15 (1833–34): 205.

94. Ibid. 2 (1820–21): 114; n.s., 1 (1834–35): 137.

95. Ibid. 11 (1829–30): 9–10; Lewis C. Gray, *History of Agriculture in the Southern United States to 1860* (Washington: The Carnegie Institution, 1933), 780.

96. Ruffin's classic work, *On Calcareous Manures*, went through at least five editions between 1832 and 1853. He also published many papers on the subject in the *American Farmer* and in the *Farmer's Register*, which he published from 1833 to 1842. For an excellent analysis of Ruffin's impact, see Craven, *Soil Exhaustion*, 134–44; also Earl Gregg Swen, *An Analysis of Ruffin's Farmer's Register, with a Bibliography of Edmund Ruffin* (Richmond: Superintendent of Public Printing, 1919).

97. *American Farmer* 2 (1820–21): 114; n.s., 1 (1834–35): 137; Ruffin, *On Calcareous Manures*, 5.

98. Ruffin, *On Calcareous Manures*, 69, 73; *American Farmer* 15 (1833–34): 154–55.

99. *American Farmer* 11 (1829–30): 9–10.

100. Ibid. s. 4, 3 (1847–48): 313–14; s. 4, 1 (1839–40): 27–28; *American Agriculturist* 1 (1842): 374; J. H. Hollander, *The Financial History of Baltimore*

(Baltimore: Johns Hopkins University Press, 1899), 60–61, 111; Wiser, "Improving Maryland's Agriculture," 112; Merritt Starr, "General Horace Capron, 1804–1885," *Journal of the Illinois State Historical Society* 18 (1925): 278.

101. The *American Farmer*, which commenced in 1819, was especially influential. The Maryland State Agricultural Society was not founded until 1848, but a predecessor existed in 1846. See Wiser, "Improving Maryland's Agriculture," 122.

102. *American Farmer* s. 4, 3 (1847–48): 313–14; s. 3, 1 (1839–40): 27–28; *American Agriculturist* 4 (1845): 43; Hollander, *Financial History of Baltimore*, 60–61, 111.

103. Wiser, "Improving Maryland's Agriculture," 112; Starr, "General Horace Capron," 278.

104. Wiser, "Improving Maryland's Agriculture," 105; Craven, *Soil Exhaustion*, 154–55.

105. Olmstead, *Journey*, 158; also see Craven, *Soil Exhaustion*, 154.

106. See Wiser, "Improving Maryland's Agriculture," 105–107, and passim, for a discussion of the importance of this pattern.

107. For patterns of guano use near cities in Maryland and Virginia, see A. B. Allen, quoted in *Cultivator* n.s., 5 (1848): 219; Edwin Bartlett, *Guano, Its Origin, Properties, and Uses* (New York: Wiley and Putnam, 1845), 29–30.

108. Edward Stabler in the *Plough, Loom, and Anvil* 1 (1849): 741; *American Farmer* 2 (1820–21): 34.

109. See *American Agriculturist* 2 (1843): 174; NYSAS, *Transactions* 2 (1842): 223–24; *American Farmer* 2 (1820–21): 34; 7 (1825–26): 9; s. 3, 2 (1840–41): 240; s. 4, 3 (1847–48): 9–13, 69; 4 (1848–49): 104.

110. *Southern Cultivator* 5 (1847): 186; *American Farmer* s. 4, 2 (1846–47): 331–32; 3 (1847–48): 9–13; *Plough, Loom, and Anvil* 1 (1849): 740–47. The pattern of substitution of guano for lime is clear in George W. Hyde, Farmbook 1840–1888, Maryland Historical Society, MS 2154; and Farm Reports, John C. Ruth Collection, Maryland Historical Society, MS 2154.

111. Gray, *Agriculture in the Southern U.S.*, 701.

112. *Southern Cultivator* 2 (1844): 86; 15 (1857): 205. In the mid-1840s, extensive use of manure was still so rare in the Southeast that the region's agricultural journals, such as the *Southern Cultivator* and *Southern Agriculturist*, seldom ran an article reporting local use or experiments. What articles they did carry on manures were almost all reprinted from Northern or European journals. The *American Farmer* 15 (1833–34): 122, reprinted an article from the *North Carolina Star* describing the remarkable agricultural improvements in Maryland and Virginia and wishing that similar improvements would occur in North Carolina.

113. As Gavin Wright notes in *The Political Economy of the Cotton South: Households, Markets, and Wealth in the Nineteenth Century* (New York: W. W. Norton, 1978), 17, cotton was not a highly exhausting crop. Much of the cotton land in the Coastal Plain and Piedmont regions of the Southeast, however, was not originally high in fertility. In the 1840s, the situation began to change and it became cheaper for many planters to restore their old lands than to move west (Ralph Betts Flanders, *Plantation Slavery in Georgia* [Chapel Hill: University of North Carolina Press, 1933], 93). This interest in land restoration triggered an interest in marl use in the early 1840s (*Southern Cultivator* 2 [1844]: 144, 147; 4 [1846]: 99), and also led

to attempts to make the best of whatever materials were locally available (*Southern Cultivator* 6 [1846]: 259). Marl and other locally available materials were so inadequate, however, that no stable fertilizer system could be established in the area until commercial fertilizers became available from outside the region.

Chapter II: The Impact of the Recycling Mentality

1. *American Agriculturist* 6 (1847): 110.
2. *Plough, Loom, and Anvil* 5 (1852–53): 304.
3. *Cultivator* 1 (1834–35): 92; n.s., 1 (1844): 114; 2 (1845): 171, 211; 7 (1850): 40, 197–98; s. 3, 7 (1859): 169; *American Farmer* 13 (1831–32): 73–74, 133; s. 4, 3 (1847–48): 6; *Working Farmer* 1 (1849): 76, 136; 2 (1850): 5, 43, 89; 4 (1852): 212; 7 (1856): 41; 11 (1859): 113; *American Agriculturist* 1 (1842): 281–82, 381; 6 (1847): 110; 19 (1860): 281; Samuel L. Dana, *A Muck Manual for Farmers* (Lowell, Mass.: Daniel Bixby, 1842), 145. See also Clarence Danhof, *Change in Agriculture: The Northern United States, 1820–1870* (Cambridge: Harvard University Press, 1969), 267; John Hannam, *The Economy of Waste Manures: A Treatise on the Nature and Use of Neglected Fertilizers* (London; Longman, Brown, Green, and Longmans, 1844), 79 and passim.
4. *Cultivator* 4 (1837–38): 101.
5. *Aberdeen Journal*, quoted in *Cultivator* 5 (1838–39): 105.
6. *Working Farmer* 11 (1859): 172.
7. American Institute of the City of New York, *Transactions* (1856): 222.
8. *Working Farmer* 11 (1859): 172; Cultivator 9 (1842–43): 162; 10 (1843–44): 108; Gouverneur Emerson, *Address Delivered before the Agricultural Society of Kent County, Delaware* (Philadelphia: National Merchant Printer, 1857), 6.
9. Horace Greeley, *Essays Designed to Elucidate the Science of Political Economy* (1870); reprint ed., New York: Arno Press, 1972), 137–38.
10. *Cultivator* n.s., 1 (1844): 364; 3 (1846): 110; U.S. Patent Office, *Agricultural Report* (1845), 1051; NYSAS, *Transactions* (1847): 376–77; *American Agriculturist* 6 (1847): 138.
11. *Cultivator* s. 3, 3 (1855): 83.
12. Ibid. 1 (1834–35): 92.
13. Ibid.; ibid. 8 (1841–42): 12; *American Farmer* 13 (1831–32): 129; s. 3, 1 (1839–40): 251.
14. See *Working Farmer* 3 (1851): 72; *American Agriculturist* 18 (1859): 378; 19 (1860): 127.
15. *National Cyclopaedia of American Biography* (Clifton, N.J.: J. T. White, 1979), 5: 135.
16. The phosphatic guano "mania" and the origins of phosphate mining are subjects of chapters 4, 5, and 8.
17. For examples see *American Agriculturist* 1 (1842): 236; *Cultivator* n.s., 4 (1847): 138; *American Farmer* 13 (1831–32): 73.
18. Justus Liebig and the Swedish chemist Jons Jacob Berzelius were often quoted. See Hannam, *Economy of Waste Manures*, 71; *Cultivator* 8 (1841–42): 110.
19. *Working Farmer* 11 (1859): 104; *Southern Cultivator* 28 (1870): 145.
20. Lemuel Shattuck, *Report of the Sanitary Commission of Massachusetts, 1850*, reprint ed. (Cambridge: Harvard University Press, 1948), 215.

21. According to Dr. Charles T. Jackson of Boston, cited in *American Agriculturist* 4 (1845): 115.

22. *Working Farmer* 2 (1850): 191.

23. *American Agriculturist* 1 (1842): 236.

24. Ibid. 3 (1844): 233.

25. MSBA, *Annual Report* 1 (1867): 233.

26. George E. Waring, *The Elements of Agriculture: A Book for Young Farmers* (Montpelier, Vt.: Samuel M. Walton, 1855), 129.

27. *Country Gentleman* 36 (1871): 451.

28. *American Agriculturist* 17 (1858): 327; *Edinburgh Journal*, condensed in *American Agriculturist* 2 (1843): 158; *British and Foreign Medico-Chirurgical Review*, quoted in Shattuck, *Report*, 215.

29. *Cultivator* n.s., 4 (1847): 128.

30. *Working Farmer* 2 (1850): 173; 1 (1849): 168; Culhbert W. Johnson, *On Fertilizers* (London: Ridgway, Piccadilly, 1839), 93–112; D. K. Minor, *Poudrette as a Manure or Fertilizer in Comparison with Other Manures* [1844?], 2–5.

31. F. A. Ismar, *American Farmer* 13 (1831–32): 73.

32. Charles T. Jackson, U.S. Patent Office, *Agricultural Report* (1856): 198–201; *American Agriculturist* 4 (1845): 115; *Working Farmer* 1 (1849): 153; 2 (1850): 191.

33. *American Agriculturist* 6 (1847): 148.

34. Ibid. 2 (1843): 152.

35. Greeley, *Essays*, 123.

36. *Working Farmer* 7 (1856): 181.

37. The best account of the history of poudrette is in *Cultivator* 8 (1841–42): 110. See also Johnson, *On Fertilizers*, 101; Hannam, *Economy of Waste Manures*, 67.

38. *Cultivator* 4 (1837–38): 112.

39. See *American Agriculturist* 2 (1843): 370; *Working Farmer* 3 (1841–42): 39.

40. *Sag Harbor Corrector*, 28 December 1839, describes the breakaway of the Lodi company from the New York company.

41. *Cultivator* 6 (1839–40): 114

42. Lodi Manufacturing Company, *An Act to Incorporate the Lodi Manufacturing Company* (New York: H. Cassidy, 1840), 4; *Cultivator* 7 (1840–41): 144.

43. *Cultivator* n.s., 1 (1844): 163.

44. Ibid. n.s., 2 (1845): 306.

45. *American Agriculturist* 22 (1863): 169. In 1863 the city paid the company $15,000 per year to remove the night soil under a five-year contract. For a description of the company and its operations, see Lodi Manufacturing Company, *An Act*, 4; Lodi Manufacturing Company, *New and Improved Poudrette* [New York, 1850?], 10; D. J. Browne, *The Field Book of Manures or the American Muck Book* (New York: A. O. Moore, 1853), 310. For an excellent later description, see *American Agriculturist* 22 (1863): 169.

46. Lodi Manufacturing Company, *An Act*, 6.

47. Lodi Manufacturing Company, *New and Improved*, 9.

48. *Working Farmer* 4 (1852): 276; NYSAS, *Transactions* (1842): 209.

49. Lodi Manufacturing Company, *An Act*, 8.

50. For examples of advertisements of the New York Poudrette Company, see *American Agriculturist* 2 (1843): 95; 3 (1844): 31; 4 (1845): 199. For advertisements

of the Lodi Manufacturing Company, see, *American Farmer* s. 4, 5 (1849–50): 291; *American Agriculturist* 15 (1856): 181; 19 (1859): 127; *Glen Cove Gazette*, 8 March 1862.

51. According to a farmer writing in the NYSAS, *Transactions* 2 (1842): 209, poudrette was theoretically twelve to fifteen times as strong as good stable manure, indicating that one or two tons would be needed per acre. Samuel W. Johnson was highly suspicious of these claims. He figured that good poudrette was worth only twice its weight in stable manure (*Essays on Peat, Muck, and Commercial Manures* [Hartford: Brown & Gross, 1859], 44–45).

52. *Cultivator* s. 3, 1 (1853): 226.

53. *American Agriculturist* 17 (1858): 95; 19 (1860): 127.

54. Ibid. 15 (1856): 181.

55. NYSAS, *Transactions* 2 (1842): 209; *Cultivator* n.s., 2 (1845): 151; 4 (1847): 126; s.3, 3 (1855): 92; *Plough, Loom, and Anvil* 1 (January 1849): 746–47; U.S. Patent Office, *Agricultural Report* (1853): 79; *American Farmer* s. 4, 4 (1848–49): 103.

56. See NYSAS, *Transactions* 2 (1842): 209.

57. S. W. Johnson, *Essays*, 44–45. For the Lodi Manufacturing Company's reaction to Mr. Johnson's analyses, see *County Gentlemen* 19 (1862): 250, 282–83.

58. *American Farmer* s. 3, 2 (1840): 165.

59. Ibid. s. 4, 9 (February, 1854): advertising pages.

60. *Cultivator* 9 (1842–43): 61; n.s., 1 (1844): 174; Lodi Manufacturing Company, *New and Improved*, 9.

61. This plant apparently did not last long. See *American Farmer* s. 4, 4 (1848–49): 266; advertisement, 5 (1849–50): 404.

62. S. W. Johnson analyzed poudrette from both the Lodi and the Liebig concerns (*Essays*, 44). For Baltimore, see *American Farmer* s. 4, 6 (1850–51): 193. For Philadelphia, see Stevenson W. Fletcher, *Pennsylvania Agriculture and Country Life, 1840–1940* (Harrisburg: Pennsylvania History and Museum Commission, 1955), 109; Edwin T. Freedley, *Philadelphia and Its Manufacturers in 1857* (Philadelphia: Edward Young, 1859), 144–45.

63. *American Farmer* s. 4, 5 (1849–50): 291.

64. *American Agriculturist* 16 (1857): 71; 17 (1858): 95.

65. Ibid. 15 (1856): 23.

66. Ibid., 181.

67. Ibid. 25 (1866): 159.

68. Ibid. 28 (1869): 153; 33 (1874): 76; 34 (1875): 79, 119.

69. Ibid. 16 (1857): 71; *American Farmer* 12 (1856–57): 292.

70. *Country Gentleman* 21 (1863): 107; *American Agriculturist* 25 (1866): 162.

71. *Bromophyte Fertilizer Co. of St. Louis, Mo.* (St. Louis: N. pub., 1871).

72. *American Farmer*, March advertisements s. 8, 6 (1877): n. p.

73. A. L. Mehring, J. Richard Adams, and K. D. Jacob, *Statistics on Fertilizers and Liming Materials in the United States*, USDA, Agricultural Research Service, Statistical Bulletin No. 191 (April 1957), indicates maximum production of 70,000 tons and then a decline to 5,000 tons in 1900. Since manufacturers stretched the term poudrette to include other materials made from slaughterhouse wastes, these figures may exaggerate the actual recycling of human excrement.

74. *American Agriculturist* 8 (1849): 153.

75. Edwin Chadwick, *Report on the Sanitary Condition of the Labouring Population of Great Britain*, 1842, ed. M. W. Flinn (Edinburgh: N. pub., 1965), 59–60, 118-19.

76. See Metropolitan Board of Works, *The Agricultural Value of the Sewage of London* (London: Edward Stanford, 1865).

77. *Scientific American* n.s., 28 (1873): 405.

78. Many reports of British proposals appeared in the American press. See *Working Farmer* 2 (1850): 11–12, 191, 205; *Cultivator* n.s., 5 (1848): 99, 205; *Scientific American* n.s., 28 (1873): 405; *Country Gentlemen* 20 (1862): 208; *American Agriculturist* 32 (1873): 91, 139; NYSAS, *Transactions* 22 (1862): 280–306.

79. *Working Farmer* 6 (1854): 99; 7 (1855): 181.

80. Ibid. 4 (1852): 275–76.

81. Ibid. 2 (1850): 206.

82. George E. Waring, *Draining for Profit and Draining for Health* (New York: Orange Judd, 1867), 227.

83. *American Agriculturist* 32 (1873): 91, 139.

84. *National Cyclopaedia of American Biography*, 6: 157; Waring, *Elements of Agriculture*, 129–31; *Working Farmer* 6 (1854–55): 246; 11 (1859–60): 49.

85. *National Cyclopaedia of American Biography*, 6: 157.

86. Horace Greeley, *What I Know of Farming* (New York: N. pub., 1871), 121–22.

87. *Scientific American* n.s., 28 (1873): 405. See also, *Country Gentleman* 38 (1873): 83.

88. MSBA, *Annual Report* (1877): 177–93.

89. *Sanitary and Fertilizer Company of the United States, Part 3, Estimated Profits from the Adoption of This System* (Philadelphia: Burk & McFetridge [1880?]), 11, 18. The profit calculations are seriously in error.

90. Gustavus A. Weber, *The Bureau of Chemistry and Soils: Its History, Activities, and Organization* (Baltimore: Johns Hopkins University Press, 1928), 24.

91. See Joel A. Tarr, "From City to Farm: Urban Wastes and the Farmer," *Agricultural History* 49 (1975): 606–609, for an extended discussion of some of the problems of recycling urban sewage.

92. NYSAS, *Transactions* 19 (1859): 34.

93. Campbell Morfit, *A Practical Treatise on Pure Fertilizers* (New York: D. Van Nostrand, 1872), 1.

Chapter III: Guano

1. K. D. Jacob, "Predecessors of Superphosphates," USDA and TVA, *Superphosphate: Its History, Chemistry, and Manufacture* (Washington: GPO, 1964), 11; R. E. Coker, "Peru's Wealth Producing Birds," *National Geographic Magazine* 37 (1920): 541.

2. Jacob, "Predecessors of Superphospates," 12; Charles L. Bartlett, *Guano: A Treatise on the History, Economy as a Manure and Modes of Applying Peruvian Guano* (Boston, 1860), 4; Amédée Francois Frézier, *A Voyage to the South-Sea, and Along the Coasts of Chile and Peru, in the Years 1712, 1713, and 1714* (London: Jonah Bowyer, 1717), 147.

3. Jorge Juan and Antonia de Ulloa, *A Voyage to South America* (1748; abridged ed., New York: Alfred A. Knopf, 1964), 219–20.

4. C. L. Bartlett, *Guano*, 4.

5. Ibid.; [Edwin Bartlett], *Guano, Its Origins, Properties, and Uses*, 2nd ed. (New York: Wiley and Putnam, 1845), 19–20.

6. *American Farmer* 6 (1824–25): 315; E. Bartlett, *Guano*, 21.

7. Abraham Rees, *The Cyclopaedia or, University Dictionary of Arts, Sciences, and Literature* (Philadelphia: Samuel F. Bradford, 1805–25), s. v. guano.

8. John Claudius Loudon, *An Encyclopaedia of Agriculture* (London: Longman, Hurst, Rees, Orne, Brown, and Green, 1825), s. v. guano; Andrew Ure, *A Dictionary of Chemistry* (Philadelphia: Desilver, 1821), s. v. guano.

9. Sir Humphrey Davy, *Elements of Agricultural Chemistry in a Course of Lectures for the Board of Agriculture* (1813; American edition, New York: Eastburn, Kirk, 1815), 263–64.

10. E. Bartlett, *Guano*, 4; *Hunt's Merchant's Magazine* 34 (1856): 118.

11. Alexander von Humboldt, *Personal Narrative of Travels to the Equinoctial Regions of the New Continent During the Years 1799–1804*, 7 vols. (London: Longmen, Hurst, Orne, and Browne, 1814–29).

12. L. Kellner, *Alexander von Humboldt* (London: Oxford University Press, 1963), 58, does not mention guano or guano islands.

13. Jacob, "Predecessors of Superphosphates," 12; George Evelyn Hutchinson, "The Biogeochemistry of Vertebrate Excretion," *Bulletin of The American Museum of Natural History* 96 (1950): 315.

14. *American Farmer* 6 (1824–25): 316–17.

15. Ducatel actually held chairs in natural history, chemistry, and geology at Baltimore's Mechanic's Institute and at the University of Maryland. (*National Cyclopaedia of American Biography* [Clifton, N.J.: J. T. White, 1979], 4: 544.

16. *American Farmer* 6 (1824–25): 446–47; *New England Farmer*, reprinted in *American Farmer* 12 (1830–31): 286; *Cultivator* 10 (1843–44): 64; Jacob, "Predecessors of Superphospate."

17. *Cultivator* 8 (1841–42): 189.

18. E. Bartlett, *Guano*, 20. For additional information on early importation to Great Britain, see Hutchinson, "Biogeochemistry of Vetebrate Excretion," 42; and Holmer J. Wheeler, *Manures and Fertilizers* (New York: Macmillan, 1914), 75–76.

19. David P. Werlich, *Peru: A Short History* (Carbondale, Ill.: Southern Illinois University Press, 1978), 79–80.

20. C. L. Bartlett, *Guano*, 3.

21. J. H. Sheppard, *A Practical Treatise on the Use of Peruvian and Ichaboe African Guano: Cheapest Manure in the World*, 2nd ed. (London: Simpka, Marshal, 1844), 5. For examples, see Sheppard, *Practical Treatise*; [Anthony Gibbs and Sons], *Guano: Its Analysis and Effects: Illustrated by the Latest Experiments* (London: W. Clowes & Sons, 1843); and anonymous, *Hints to Farmers on the Nature, Purchase, and Application of Peruvian, Bolivian, and African Guano* (London: Longman, 1845). See E. Bartlett, *Guano*, for references and quotations from many of the British sources.

22. *Cultivator* 8 (1841–42): 189; 9 (1842–43): 156. See also *American Farmer* s. 3, 3 (1841–42): 349. Ruffin's *Farmer's Register* carried at least five accounts of guano in Great Britain in 1841 and 1842.

23. Samuel L. Dana, *A Muck Manual for Farmers* (Lowell, Mass.: Daniel Bixby, 1842), 145–47. The analysis was attributed to Voelckel.

24. *American Agriculturist* 3 (1844): 99.

25. James E. Teschemacher, *Essay on Guano* (Boston: A. D. Phelps, 1845), 4; Teschemacher speech of October 1843, *American Agriculturist* 3 (1844): 24. The *Farmer's Magazine* (London), n.s., 13 (1846): 422–36, reprinted Teschemacher's *Essay*, indicating that his work was also of interest in Great Britain. For reports of other experiments between 1842 and 1844, see Solon Robinson, *Guano, A Treatise of Practical Information for Farmers* (New York, 1852), 94; E. Bartlett, *Guano*, 30–31, 88–91; *Plough, Loom, and Anvil* 1 (1848): 102; *Farmer's Cabinet* 8 (1843): 135, quoted in Stevenson Whitcomb Fletcher, *Pennsylvania Agriculture and Country Life, 1840–1940* (Harrisburg: Pennsylvania Historical and Museum Commission, 1955), 107–108.

26. E. Bartlett, *Guano*, 28, 83.

27. Ibid., 28; Frank Roy Rutter, *South American Trade of Baltimore* (Baltimore: Johns Hopkins Press, 1897), 42.

28. E. Bartlett, *Guano*.

29. Ibid., 83; Pennsylvania State Agricultural Society, *Transactions* 1 (1845): 182.

30. F. M. L. Thompson, "The Second Agricultural Revolution, 1815–1880," *Economic History Review* s. 2, 21 (1968): 70.

31. E. Bartlett, *Guano*, 18.

32. John Collis Nesbit, *On Peruvian Guano: Its History, Composition, and Fertilizing Qualities* (London: Longman, 1852), 4, 8; Daniel Lee, "Treatise on the Relation of Peruvian Guano to American Agriculture," *Peruvian Guano Trade* (Washington: W. H. More, 1854), 24, 26; *Cultivator* n.s., 1 (1844): 182.

33. *American Agriculturist* 4 (1845): 43.

34. Ibid. 3 (1844): 99.

35. *Cultivator* n.s., 2 (1845): 45, 340; 3 (1846): 19.

36. Ibid. n.s., 4 (1847): 160.

37. *Agriculator*, quoted in *Cultivator* s. 3, 1 (1853): 226.

38. *Cultivator* n.s., 5 (1848): 219.

39. Ibid. s. 3, 1 (1853): 226; *Southern Cultivator* 9 (1851): 35.

40. T. S. Pleasants to J. S. Skinner, 21 August 1844, printed in E. Bartlett, *Guano*, 29–30; *American Farmer* s. 4, 5 (1849–50): 189.

41. Quoted in E. Bartlett, *Guano*, 31.

42. *American Farmer* s. 4, 1 (1845–46): 69–70; 2 (1846–47): 334–45.

43. J. E. Teschemacher, quoted in *American Agriculturist* 4 (1845): 156.

44. *American Farmer* s. 4, 6 (1850–51): 101, 174, 193; *Cultivator* n.s., 5 (1848): 188.

45. *American Agriculturist* 8 (1849): 153; 11 (1853–54): 329; *American Farmer* s. 3, 3 (1841–42): 349; *Cultivator* 8 (1841–42): 189.

46. E. Bartlett, *Guano*, 30–31; *Plough, Loom, and Anvil* 1 (1848–49): 743. For a list of farmers reporting early experiments, see Vivian Wiser, "Improving Maryland's Agriculture: 1840–1860," *Maryland Historical Magazine* 62 (1962): 109.

47. *Working Farmer* 2 (1850): 38; 3 (1851): 114.

48. *Hunt's* 36 (1857): 638; *American Agriculturist* 3 (1844): 220.

49. J. S. Skinner speech (1844?), quoted in E. Bartlett, *Guano*, 29. See also *Hunt's* 34 (1856): 118; *Working Farmer* 4 (1852): 159.

50. *Working Farmer* 6 (1854): 145; 4 (1852): 329.

51. Daniel Lee, in *Southern Cultivator* 11 (1853): 329.

52. Daniel Lee developed this concept in his "Treatise on Peruvian Guano," 22. Also see Charles T. Jackson, U.S. Patent Office, *Agricultural Report* (1854): 102; Teschemacher, *Essay on Guano*, 13.

53. *American Agriculturist* 4 (1845): 130.

54. U.S. Treasury Department, *Annual Report and Statement on the Commerce and Navigation of the United States*, 1848, 1849, 1855. Figures are for fiscal year ending in June. The 1845 figures are from E. Bartlett, *Guano*, 83.

55. The *Cultivator* peaked with ten articles in 1844, followed by seven in 1845 and eight in 1846. The *Southern Planter* had none in 1842, four in 1845, and nineteen in 1850.

56. Clarence Danhof, *Change in Agriculture: The Northern United States, 1820–1870*, (Cambridge: Harvard University Press, 1969), 280-87, provides a general model for the spread of agricultural innovation. He divides farmers into four groups: a small group of innovators, a small group of imitators who follow the lead of the innovators, much larger groups of gradualists who change very slowly, and traditionalists who resist all change. Danhof, however, seems to underestimate the rate at which an innovation such as guano could spread.

57. *American Farmer* s. 4, 1 (1845–46): n.p.; 2 (1846–47): 334–35.

58. Ibid. s. 4, 3 (1847–48): 10.

59. Quoted in ibid., 115.

60. Ibid.

61. U.S. Patent Office, *Agricultural Report* (1848): 469.

62. Ibid. (1853): 81; *American Farmer* s. 4, 8 (1852–53): 41–43.

63. *American Farmer* s. 4, 8 (1852–53): 41–43.

64. *American Farmer* s. 4, 8 (1852–53): 42, 103. Also see *American Agriculturist* 9 (1850): 202; *Southern Planter* 10 (1850): 118.

65. *Cultivator* s. 3, 1 (1853): 354.

66. *American Agriculturist* 8 (1849): 23.

67. *American Farmer* s. 8, 9 (1880): 169.

68. A typical Peruvian guano had a composition of 11 to 16 percent nitrogen, 8 to 12 percent phosphorus, and 2 to 3 percent potassium (Gilbeart H. Collings, *Commercial Fertilizers: Their Sources and Use* [Philadelphia: P. Blakiston's Son, 1938], 121).

69. *Southern Planter* 5 (1845): 140.

70. See the reports of experiments by J. A. Pearce of Kent County, Maryland, and Edward Stabler of Montgomery County, Maryland, *American Farmer* s. 4, 1 (1845–46): 51, 69.

71. For examples of failures, see *Southern Planter* 5 (1845): 140; *American Agriculturist* 5 (1846): 77–78; NYSAS, *Transactions* 5 (1845): 637.

72. *Southern Planter* 5 (1845): 59; *American Agriculturist* 5 (1846): 196; 8 (1849): 163.

73. For examples, see E. Bartlett, *Guano*, 92; *DeBow's Review* 16 (1854): 463.

74. For example, see C. L. Bartlett, *Guano*, 7, 9. The sections on experience were generally longer.

75. Robinson, *Guano*, 36.

76. Ibid., 33–34; *American Agriculturist* 12 (1854): 208.

77. Lee, "Treatise on Peruvian Guano," 23.

78. Gavin Wright, *The Political Economy of the Cotton South: Households, Markets, and Wealth in the Nineteenth Century* (New York: W. W. Norton, 1978), 17, 19–22. Wright is undoubtedly correct in his assertion that cotton is not a highly exhaustive crop. Almost any field crop grown without the application of manure or fertilizer, however, will tend to reduce the natural fertility of the soil.

79. Rosser H. Taylor, "Fertilizers and Farming in the Southeast, 1840–1950, Part 1, 1840–1900," *North Carolina Historical Review* 30 (1953): 305–306.

80. Taylor, "Fertilizers in the Southeast," 308–307.

81. *Southern Cultivator* 6 (1846): 32–39; Robert Russell, *North America: Its Agriculture and Climate* (Edinburgh: Adam & Charles Black, 1857), 165.

82. The *Southern Cultivator* published no articles on guano in 1846, one in 1847, and two in 1848. *Southern Cultivator* 11 (1853): 292, reported that only in Maryland, Delaware, and Virginia did many farmers use guano.

83. *Southern Cultivator* 11 (1853): 6, 45, 136; 17 (1859): 360.

84. Ibid. 17 (1859): 255.

85. Quoted in ibid. 18 (1860): 81.

86. *Hunt's* 43 (1860): 259.

87. The best firsthand descriptions of the guano islands during the height of the guano trade are: Anonymous letter (20 July 1853) published in *Liverpool Albion*, reprinted in *Country Gentleman* 2 (1853): 279. Anonymous letter published in *New York Evening Post* reprinted in *Country Gentleman* 8 (1856): 19. A. J. Boyd, "Reminiscence of the Chincha Islands," *Queensland Geographical Journal* 8 (1893): 3–12. A private letter of 19 February 1856, published in the *Boston Traveler*, described a visit to the islands. It was reprinted in *American Agriculturist* 12 (1854): 84 and *Working Farmer* 6 (1857): 85–86. The account of John R. Congdon, who visited the Chincha Islands on the bark *Thornton* in August 1853, was published in the *Providence Journal* and reprinted in *American Agriculturist* 11 (1853–54): 99–100. Frederic A. Lucas, who was later a director of the American Museum of Natural History, visited the islands in the fall of 1869. This account is published in Robert C. Murphy, *Bird Islands of Peru* (New York, G. P. Putnam's Sons, 1925), 98–108. Maurice F. Nash, of New York, visited the islands in the fall and summer of 1855. Accounts of his trip appeared in *American Institute*, (1856): 221–29, and were reprinted in *Plough, Loom, and Anvil* (1857–58): 69–76. *The National Magazine* 3 (1853): 553–56, had an anonymous description of the islands. George Washington Peck visited the Chincha Islands in November 1853. A letter describing his visit was published in the *New York Times* and reprinted in *Working Farmer* 6 (1854): 17–19. Peck published an extended version of his observations in *Melbourne and the Chincha Islands* (New York: C. Scribner, 1954), 138–225. Hutchinson, "The Biogeochemistry of Vertebrate Excretion," 33–43, contains the best recent discussion of the Chincha Islands and other guano islands along the Peruvian coast.

88. Murphy, *Bird Islands*, 60, 65; *American Agriculturist* 6 (1847): 123.

89. Coker, "Peru's Wealth Producing Birds," 552.

90. Robert Cushman Murphy and Grace E. Barstow Murphy, "Peru Profits from Sea Fowl," *National Geographic Magazine* 145 (1959): 395–413; Boyd, "Reminiscence," 5; Teschemacher, *Essay on Guano*, 5.

91. For a detailed account of the mechanics of the guano operations and trade, and the horrible conditions on the islands, see Richard A. Wines, "From Recycled Wastes to Commercial Fertilizers: The Evolution of a Technological System in the Eastern United States, 1800–1880" (Ph.D. diss., Brown University, 1981), 108–14.

92. *De Bow's* 19 (1855): 220–21.

93. Coker, "Peru's Wealth Producing Birds," 543.

94. Werlich, *Peru*, 98.

95. Ibid.

96. Ibid., 81.

97. Rutter, *South American Trade*, 40–41.

98. J. Randolph Clay to Marcy, 26 March 1856, in U.S. Congress, 35: 2, *Senate Executive Document* No. 25, 33–35.

99. Werlich, *Peru*, 80–81; Anthony Gibbs and Sons, *Peruvian and Bolivian Guano: Its Nature, Properties, and Results* (London: James Ridgway, 1844), 3.

100. Rutter, *South American Trade*, 41; U.S. Congress, 35: 2 *Senate Executive Document* No. 25, 31–32.

101. Roy F. Nichols, *Advance Agents of American Destiny* (Philadelphia: University of Pennsylvania Press, 1956), 161.

102. Ibid.; *American Farmer* s. 4, 8 (1852–53): 18.

103. E. Bartlett, *Guano*, n.p.

104. Rutter, *South American Trade*, 41.

105. Ibid., 46; Hamilton Owens, *Baltimore on the Chesapeake* (Garden City, N.Y.: Doubleday, Doran, 1941), 252–56.

106. Rutter, *South American Trade*, 46. The Barredas were also involved in other trades, as is documented in F. Barreda and Brother Co. Records, MS 2104, Maryland Historical Society.

107. Rutter, *South American Trade*, 48.

108. *Southern Cultivator* 9 (1851): 71; U.S. Congress, 31: 1, *Senate Executive Document* No. 80, 1–2.

109. Peck, *Chincha Islands*, 211.

110. Murphy, *Bird Islands*, 101, 107; T. Courtenay Whedbee, *The Port of Baltimore in the Making 1828 to 1878* (Baltimore: F. Bowie Smith & Son, 1953), 56.

111. Advertisement, *American Farmer* 9 (1854): n.p.; *Southern Cultivator* 9 (1851): 107.

112. Advertisement, *American Farmer* (1846): 294, 319; *American Agriculturist* 4 (1845): 294; 6 (1847); 134.

113. Advertisement, *American Farmer* 12 (1856–57): n.p.

114. *American Farmer* s. 4, 8 (1852–53): 42; *American Agriculturist* 9 (1850): 321; *Working Farmer* 2 (1850): 181.

115. F. Barreda and Brother, in U.S. Congress, 35: 2, *Senate Executive Document* No. 25, 31–32.

116. *Glen Cove Gazette*, 4 July 1868, n.p.

117. U.S. Treasury Department, *Commerce and Navigation* (1856–58).

118. Rutter, *South American Trade*, 42–43.

119. Owens, *Baltimore*, 254–55.

120. See R. H. Taylor, "Commercial Fertilizer in South Carolina," *The South Atlantic Quarterly* 29 (1930): 182.

121. *American Farmer* s. 4, 5 (1849–58): 189.

122. Ibid. 13 (1857–58): 35, 39.

123. *American Agriculturist* 10 (1851): 208; *American Farmer* 9 (1853–54): 233; U.S. Patent Office, *Agricultural Report* (1851), 2.

124. *Southern Planter* 10 (1850): 318.

125. C. L. Bartlett, *Guano*, 10–11; *Working Farmer* 1 (1849): 57.

126. Robinson, *Guano*, 35–36.

127. E. Bartlett, *Guano*, n.p.; *American Agriculturist* 4 (1845): 108–109.

128. *Cultivator* s. 3, 5 (1857): 142; NYSAS, *Transactions* (1846): 634; 10 (1850): 540; *American Agriculturist* 3 (1844): 34.

129. *American Farmer* s. 4, 14 (1858–59): 17; 15 (1859–60): 115; *American Agriculturist* 17 (1858): 95; U.S. Congress, 35: 2, *Senate Executive Document* No. 25, 70.

130. *American Agriculturist* 31 (1872): 209.

131. *American Farmer* s. 4, 6 (1850–51): 19; 12 (1856–57): 323; U.S. Patent Office, *Agricultural Report* (1853), 81; *Glen Cove Gazette*, 11 November 1858, n.p.

132. *American Institute* (1858): 343; *American Agriculturist* 28 (1869): 283.

133. Nesbit, *On Peruvian Guano*, 28; *American Agriculturist* 8 (1849): 125. This problem plagued British farmers from the beginning of guano importation there. See *American Agriculturist* 8 (1849): 125; 9 (1850): 122, 228; 10 (1851): 36; *Working Farmer* 3 (1851): 136; *Cultivator* n.s., 1 (1849): 245, 364; 7 (1850): 198; s. 3, 2 (1854): 227, 276, 371. For reports of fraud in the United States see *American Farmer* s. 4, 10 (1854–55): 368–69; 13 (1857–58): 17; *Cultivator* s. 3, 4 (1856): 66; *American Agriculturist* 5 (1855): 164–65.

134. *American Institute* (1867): 451.

135. Nesbit, *On Peruvian Guano*, 28.

136. See advertisement of Griffing Brothers & Company, *American Agriculture* 19 (1860): 127, which warns farmers about this practice.

137. *Delaware State Journal*, quoted in *American Farmer* s. 4, 8 (1852–53): 18.

138. Marcy to Clay, 30 August 1853, U.S. Congress, 33: 1, *House Executive Document* No. 70, 2–3.

139. U.S. Congress, 33: 1, *House Report* No. 347, 7.

140. Edward Stabler letter to *American Farmer* 6 (1850–51): 141; *Plough, Loom, and Anvil* 2 (1849–50): 787.

141. Wiser, "Improving Maryland's Agriculture," 110.

142. Walsh to Clayton, 30 May 1850, U.S. Congress 31: 1, *Senate Executive Document* No. 59, 7–8.

143. Clayton to Clay, 26 April 1850; Clay to Clayton, 12 August 1850; Clayton to Tirado, 7 June 1850; Tirado to Clay, 10 June 1850, *Senate Executive Document* No. 59, 6–7, 1, 2, 3, 5.

144. James Richardson, *A Compilation of the Messages and Papers of the Presidents, 1789–1897* (Washington: GPO, 1897), 5: 83.

145. *American Agriculturist* 11 (1853–54): 168; Marcy to Clay, 30 August 1853, U.S. Congress, 33: 1, *House Executive Document* No. 70, 23; Clay to Marcy, 11 November 1853, U.S. Congress, 33: 1, *House Executive Document* No. 70, 4.

146. U.S. Congress, 33: 1, *Senate Miscellaneous Document* No. 18, 1 and *House Report* No. 347, 1–2.

147. Clay to Marcy, 11 January 1854, U.S. Congress, 35: 2, *Senate Executive Document* No. 25, 2.

148. Clay to Marcy, 24 August 1854, U.S. Congress, 35: 2, *Senate Executive Document* No. 25, 5–10.

149. U.S. Congress, *Congressional Globe* 33: 1 (31 July 1854): 2024; (1 August 1854): 2040.

150. *De Bow's* 18 (1855): 33.

151. *American Farmer* s. 4, 11 (1855–56); *De Bow's* 20 (1856): 745.

152. *De Bow's* 25 (1858): 147; Wiser, "Improving Maryland's Agriculture," 111.

153. Peck, *Chincha Islands*, 272–84.

154. Paul W. Gates, *The Farmer's Age: Agriculture, 1815–1860* (New York: Holt, Rinehart, and Winston, 1960), 321; Rutter, *South American Trade*, 96.

155. Fletcher P. Veitch, "Maryland's Early Fertilizer Laws and Her 1st State Agricultural Chemist," *Journal of the Association of Official Agricultural Chemists* 17 (1934): 476–77.

156. Ibid.; Wiser, "Improving Maryland's Agriculture," 110, 126.

157. Rutter, *South American Trade*, 45; *Hunt's* 31 (1854): 232; *American Farmer* 10 (1854–55): 59; Veitch, "Maryland's Fertilizer Laws," 477.

158. NYSAS, *Transactions* (1847): 369.

159. National Fertilizer Association, *The Significance of the Word "Guano" in Fertilizer Terminology*, prepared by Charles J. Brand (Washington: National Fertilizer Association, 1931), passim.

Chapter IV: The Guano Island Mania

1. See John Basset Moore, *Digest of International Law*, 8 vols. (Washington: GPO, 1906), 1: 567–68, for a list of the seventy islands formally claimed. Some of these islands, however, were duplicates, nonexistant, later removed from the list, or never officially recognized.

2. *Cultivator* n.s., 1 (1844): 182; *American Agriculturist* 3 (1844): 188.

3. Ibid. n.s., 1 (1844): 245.

4. *American Agriculturist* 2 (1843): 171.

5. *Cultivator* n.s., 1 (1844): 182.

6. *American Agriculturist* 3 (1844): 7.

7. See George Lawrence Green, *Panther Head: The Full Story of the Bird Islands Off the Southern Coasts of Africa, the Men of the Islands, and the Birds in Their Millions* (London: S. Paul, 1955), 86–99, for an excellent description of this fantastic episode. For a contemporary American account, see *American Agriculturalist* 3 (1844): 220. For sample advertisements, see *American Agriculturalist* 3 (1844): 287 and 6 (1847): 134. For the best modern description of the islands, see George Evelyn Hutchinson, "The Biogeochemistry of Vertebrate Excretion," *Bulletin of the American Museum of Natural History* 96 (1950): 146–47.

8. *American Agriculturist* 3 (1844): 287. On 16 August 1844, D. P. Gardner, a lecturer on "Agriculture and Analysis" at the New York Agricultural Institute, reported the recent arrival of "large quantities" of African guano in New York (*Cultivator* n.s., 1 [1844]: 291).

9. *Cultivator* s. 3, 4 (1856): 133.

10. Ibid. n.s., 1 (1844): 238, acknowledged receiving a sample of African guano.

11. Edwin Bartlett, *Guano, Its Origin, Properties, and Uses* (New York: Wiley and Putnam, 1845), 28; Frank Roy Rutter, *South American Trade of Baltimore* (Baltimore: Johns Hopkins University Press, 1897), 42.

12. James E. Teschemacher, *Essay on Guano* (Boston: A. D. Phelps, 1845), 18; *Cultivator* n.s., 2 (1845): 140; Solon Robinson, *Guano: A Treatise of Practical Information for Farmers* (New York, 1852), 82; U.S. Patent Office, *Agricultural Report* (1854): 93–96; D. J. Browne, *The Field Book of Manures or the American Muck Book* (New York: A. O. Moore, 1853), 284; John Collis Nesbit, *On Peruvian Guano: Its History, Composition, and Fertilizing Qualities* (London: Longman, 1852), 24.

13. Teschemacher, *Essay*, 18; *Cultivator* n.s., 2 (1845): 171.

14. *Farmer's Magazine* (London), quoted in *American Institute* (1851): 533. For similar arguments, see Andrew Ure, *A Dictionary of Arts, Manufacturing, and Mines* (New York: D. Appleton, 1847), 533. For a fuller description of the Lobos affair, see Richard A. Wines, "From Recycled Wastes to Commercial Fertilizers: The Evolution of a Technological System in the Eastern United States, 1800–1880," (Ph.D. diss., Brown University, 1981), 151.

15. Gilbeart H. Collings, *Commercial Fertilizers: Their Sources and Use* (Philadelphia: P. Blakiston's Son, 1938), 123–24; R. E. Coker, "Peru's Wealth Producing Birds," *National Geographic Magazine* 37 (1920): 55.

16. *New York Times*, 30 April 1878, 8.

17. In 1853 the consignees were chartering transient vessels at $20.00 per ton. See U.S. Congress 34: 3, *Senate Report* No. 397, 14.

18. *Times* (London), 1 May 1852, 5; 20 May 1852, 6; 2 June 1852, 3; 26 June 1852, 8; U.S. Congress, 34: 3, *Senate Report* No. 397 (18 February 1857), 245; Thomas Wentworth Buller, *Remarks on the Monopoly of Guano* (London: James Ridgway, 1852), 14–19, 27–39.

19. Jewett to Webster, 2 June 1852, U.S. Congress, 34: 3, *Senate Report* No. 397, 1.

20. Webster to Jewett, 5 June 1852, U.S. Congress, 34: 3, *Senate Report* No. 397, 2.

21. See *New York Times*, 6 November 1852, 1; 8 November 1852, 5; and 9 November 1852, 4–5, for a series of articles and letters related to charges by William Cullen Bryant, editor of the *Post*, that Webster was engaged in a speculative scheme with Benson.

22. U.S. Congress, 34: 3, *Senate Report* No. 397, 203, 246.

23. Juan y de Osma to Webster, 25 June 1852, U.S. Congress, 34: 3, *Senate Report* No. 397, 30–32.

24. Ibid., 5. The Lobos affair, however, did not end here. Benson's claims for damages dragged on for years. See Roy F. Nichols, *Advance Agents of American Destiny* (Philadelphia: University of Pennsylvania Press, 1956), 162–70, for a more detailed account of the diplomatic aspects of the affair.

25. For details see Nichols, *Advance Agents*, 170–74, and L. Gruss, "The 'Mission' to Ecuador of Judah P. Benjamin," *Louisiana Historical Quarterly* 23 (1940): 162–65. See also U.S. Patent Office, *Agricultural Report* (1854): 95, and Hutchinson, "Biogeochemistry of Vertebrate Excretion," 121.

26. U.S. Patent Office, _Agricultural Report_ (1854), 96; advertisement, _American Farmer_ 9 (February, 1854) n.p. It is not always possible to determine where this guano came from. The guano operators tried to keep their sources secret to avoid competition and to prevent interference by the Mexican government. Sometimes they used the term "Mexican guano" generically to describe any soft, low analysis guano from the Caribbean area.

27. For a good description of the process of formation for phosphatic guanos (which included the Mexican variety), see Hutchinson, "Biogeochemistry of Vertrebrate Excretion," 88.

28. _Working Farmer_ 6 (1854): 63.

29. William J. Taylor, "Investigation on the Rock Guano from the Islands of the Caribbean Sea," _Proceedings of the Academy of Natural Sciences of Philadelphia_ (March, 1857): 91–97.

30. _American Journal of Science_ 22 (1856): 299; 36 (1863): 423–24.

31. Campbell Morfit, "On Columbian Guano," _Journal of the Franklin Institute_ 60 (1855): 325.

32. Ibid., 329.

33. Charles U. Shepard, _American Journal of Science_ 22 (1856): 96.

34. A. A. Hayes, _American Journal of Science_ 22 (1856): 300.

35. In addition to Morfit, Shepard, and Hayes cited above, see items by A. Snowden Piggot, _American Journal of Science_ 22 (1956): 299–300; 23 (1857): 120–21; and James Higgins and Charles Bickell, 23 (1857): 121–23. Higgins, Bickell, and Piggot all disagree with Shepard's theories. Some aspects of this debate also appeared in _American Farmer_ s. 4, 11 (1855–56): 89, 109, 376–77; 13 (1857–58): 81, 181–82, 284, 337. These scientists also debated whether the phosphates were soluble, making the Colombian guano a "natural superphosphate."

36. _American Farmer_ 13 (1857–58): 79–80.

37. U.S. Congress, 34: 3, _Senate Executive Document_ Nos. 25, 35.

38. John Collis Nesbit, _The History and Properties of the Different Varieties of Natural Guanos_ (London: Rogerson & Tuxford, 1859), 42.

39. Nichols, _Advance Agents_, 203, provides information on further diplomatic developments in the case.

40. After the _Country Gentleman_ exposed this deception, Shelton had the temerity to admit that one of the main reasons he added the Peruvian guano was to give his product the smell of ammonia which many farmers considered evidence of a strong fertilizer. Shelton later claimed that he used "Mexican" guano to produce "Chilian" guano, but he may have been using the term generically. The entire episode resulted in considerable editorial feuding between the _Country Gentleman_ and Mapes's _Working Farmer_. Mapes maintained that he performed only the role of a "miller" and that, at any rate, the "Chilian guano" he produced was actually a valuable fertilizer. For articles on the Chilian guano fraud, see _Country Gentleman_ 5 (1855): 117–18, 296, 343–44, 360, 405; 6 (1855): 25, 76, 93, 122; _Working Farmer_ 7 (1855): 126–27, 169; _American Farmer_ s. 4, 11 (1855–56): 22, 49.

41. U.S. Congress, 34: 3, _Senate Executive Document_ Nos. 25, 40; _Hunt's Merchant's Magazine_ 34 (1856): 437.

42. Nichols, _Advance Agents_, 185; _Hunt's_ 34 (1856): 435.

43. U.S. Congress, 34: 3, _Senate Executive Document_ No. 25, 93–95.

44. Edwin H. Bryan, Jr., *American Polynesia and the Hawaiian Chain* (Honolulu: Tongg Publishing, 1942), 27.

45. U.S. Congress, 23: 2, *House Executive Document* No. 105.

46. William Stanton, *The Great U.S. Exploring Expedition of 1838–1842* (Berkeley: University of California Press, 1975), 217–34.

47. Ibid., 7.

48. U.S. Congress, 34: 1, *Senate Miscellaneous Document* No. 60, 8, 9.

49. American Guano Company, *Report to the Stockholders of the American Guano Company* (Brooklyn, N.Y.: Jacob & Brockway, 1857), 7–8.

50. Ibid., 21.

51. American Guano Company, *Prospectus of the American Guano Company* (New York: J. F. Trow, 1855), 4–5, 7.

52. American Guano Company, *Report to the Stockholders*, 6.

53. Ibid., 9–12.

54. American Guano Company, *Prospectus*, 10–11; American Institute *Transactions* (1855): 258, 414–15.

55. U.S. Congress, 34: 1, *Congressional Globe*, 16 April 1856, 921; 22 May 1856 1297; 20 May 1856, 1299; *Senate Executive Document* No. 60. For a detailed discussion of the debates over the Guano Island Bill, see Wines, "From Recycled Wastes," 169–73.

56. See Glyndon G. Van Deusen, *William Henry Seward* (New York: Oxford University Press, 1967), 147.

57. U.S. Congress, *Congressional Globe*, 22 July 1856, 1698; 24 July 1856, 1741; 22 July 1856, 1696–97, contains a copy of the proposed bill, S. 339.

58. U.S. Congress, 36: 1 *Congressional Globe*, 30 March 1860, 1426.

59. American Guano Company, *Report to the Stockholders*, 34. The company had actually been organized the previous September.

60. Ibid., 9–11, 25–26.

61. Quoted in ibid., 28–31.

62. Ibid., 13; U.S. Congress, 35: 1, *Senate Report* No. 307, 2–5; also see Nichols, *Advance Agents*, 191.

63. U.S. Congress 35: 1, *Senate Report*, No. 307, 8.

64. *American Farmer* 12 (1856–57): 381–87; *Plough, Loom, and Anvil* 10 (1857): 148.

65. Bryan, *American Polynesia*, 135.

66. For details, see Nichols, *Advance Agents*, 190, 201.

67. See advertisement, *American Agriculturist* 18 (1859): 63.

68. The best descriptions of guano operations on these islands are Bryan, *American Polynesia*, 44; James D. Hague, "On the Phosphatic Guano Islands of the Pacific Ocean," *American Journal of Science* 34 (1862): 224–43: Mabel H. Closson, "Under the Southern Cross," *Overland Monthly* 21 (1893): 205–16.

69. *Dictionary of American Biography* (New York: C. Scribner's, 1964), 19: 279; see advertisement, *American Agriculturist* 19 (1860): 30.

70. Rossiter W. Raymond, "Biographical Notice of James Duncan Hague," *Transactions of the American Institute of Mining Engineers* (1908), 34. For problems with the first cargoes of American guano, see *American Farmer* s. 5, 1 (1859–60): 15, 112–13, and Hague, "Phosphatic Guano Islands," 237. Hague's scientific paper describing the guano on the Pacific Islands was based on chemical analyses done at

Yale's Shefield Laboratory where Professors George J. Brush and Samuel W. Johnson were his friends, and his brother, Arnold, was a student.

71. United States Guano Company, *Report to the Stockholders*, (New York, 1859): 7–12. Christmas Island had been discovered by Captain James Cook in 1777, but Benson located a Captain John Stetson of New Haven who claimed to have discovered guano deposits there during a shipwreck.

72. Nichols, *Advance Agents*, 196.

73. Ibid., 197; U.S. Guano Company, *Report*, 17.

74. See Nichols, *Advance Agents*, 197, and Bryan, *American Polynesia*, 39–40, for descriptions of this conflict.

75. Moore, *Digest of International Law* 1: 567–68.

76. A list of forty-eight islands claimed up to 1859 appeared in the New York *Tribune* and was reprinted in *Hunt's* 41 (1859): 476. This list includes Benson's forty-four islands; at least eleven of these do not exist. See Hague, "Phosphatic Guano Islands," 19, for a critique of the list.

77. Ibid.

78. Bryan, *American Polynesia*, 30, 50, 67–68; for advertisements, see *Country Gentleman* 13 (1865): passim.

79. For details, see Nichols, *Advance Agents*, 192–94.

80. Ibid., 194–95; Bryan, *American Polynesia*, 192.

81. For the best description of Ocean and Nauru, see Hutchinson, "Biogeochemistry of Vertebrate Excretion," 213–22. See also G. A. Pitman, *Nauru: The Phosphate Island* (London: Longmans, 1959),8.

82. *Hunt's* 41 (1859): 476.

83. Bryan, *American Polynesia*, 30, 61, 129, 132; Albert F. Ellis, *Adventuring in Coral Seas* (Sidney, Australia: Augus & Robertson, 1936), 18.

84. U.S. Congress 36: 1, *Congressional Globe* 30 March 1860, 1425.

85. Moore, *Digest of International Law* 1: 576, 579; Nichols, *Advance Agents*, 186–87, 204. The State Department listed Sombrero in 1867 as an American guano island, but in 1868 Secretary of State Seward could find no evidence that American title had ever been recognized by the President. For an excellent description of guano operations on Sombrero, see *American Farmer* (1860–61): 68.

86. Augustus Voelcker, *On Phosphatic Guanos* (London: W. Clowes & Sons, 1876), 30.

87. Nichols, *Advance Agents*, 186–87; *National Cyclopaedia of American Biography* (Clifton, N.J.: J. T. White, 1979), 7: 178. For advertisements, see *American Agriculturist* 18 (1859): 255, 386; 19 (1860): 30.

88. Pacific Guano Company, *The Pacific Guano Company: Its History; Its Products and Trade; Its Relation to Agriculture* (Cambridge: Riverside Press, 1876), 24–30; Green H. Hackworth, *Digest of International Law* (Washington: GPO, 1940–44) 1: 516. See Donald Glidden, "The Story of Swan Island," in Pacific Guano Company Collection, Baker Library, Harvard University Business School, for an informal account of conditions on the Swan Islands.

89. U.S. Congress, 36: 1 *Senate Executive Document* No. 37, 5, 8; Navassa Phosphate Company, *Navassa Phosphate Company* (Baltimore: N. pub., 1864), 6.

90. *American Farmer* 13 (1857–58): 325. For information on the founding of the company, see *Navassa Phosphate Company*.

91. See Eugene Gaussoin, *Memoir on the Island of Navassa (West Indies)* (Balti-

more: N. pub., 1866), for a description of the island during Gaussoin's six-week visit.

92. T. Courtenay Whedbee, *The Port of Baltimore in the Making 1828 to 1878* (Baltimore: F. Bowie Smith & Son, 1953), 55–56.

93. U.S. Congress, 36: 1, *Senate Executive Document* No. 37, 16–23.

94. U.S. Congress 35: 2, *Congressional Globe*, 575.

95. U.S. Congress, 36: 1, *Senate Report* No. 280, 15.

96. Ibid., 16.

97. Nichols, *Advance Agents*, 203.

98. For documents on this case, see U.S. Congress, 40: 2, *Senate Executive Document* No. 38, and U.S. Congress, 40: 3, *House Miscellaneous Document* No. 10. Also see Nichols, *Advance Agents*, 203–204.

99. *Times* (London) and *Farmer's Magazine* (London), quoted in *Working Farmer* 5 (1853–54): 43. See Hutchinson, "Biogeochemistry of Vertebrate Excretion," 154, 273.

100. Nesbit, *Natural Guanos*, 43–44; *American Farmer* s. 4, 13 (1857–58): 234. See Hutchinson, "Biogeochemistry of Vertebrate Excretion," 308.

101. *American Agriculturist* 18 (1858): 255; Nesbit, *Natural Guanos*, 36–37; Voelcker, *On Phosphatic Guanos*, 12, 14.

102. Augustus Voelcker, "On the Chemical Composition of Phosphatic Minerals Used for Agricultural Purposes," (London: William Clowes & Sons, 1875), 14–15, 32–33, 35; George W. Howard, *The Monumental City, Its Past History and Present Resources* (Baltimore: J. D. Ehlers, 1873), 236.

103. For a list of these islands, see Moore, *Digest of International Law*, 567–80. Most of these claims were filed by James W. Jewett in 1868–69 and 1879–80. Also see Nichols, *Advance Agents*, 204–205.

104. See Bryan, *American Polynesia*, 186; Hutchinson, "Biogeochemistry of Vertebrate Excretion," 198; and Nichols, *Advance Agents*, 206.

105. Nichols, *Advance Agents*.

106. See U.S. Treasury Department, *Annual Report and Statement on the Commerce and Navigation of the United States* from 1848 to 1902, for statistics on guano imports. Since guano was a duty-free import, there was little incentive for accurate record-keeping. The figures for imports from the bonded guano islands after 1870 appear especially carelessly assembled. Probably the amounts were significantly underreported. Moreover, the valuations are generally highly unrealistic.

Chapter V: The Impact of Phosphatic Guano

1. Andrew Ure, *A Dictionary of Arts, Manufactures, and Mines* (New York: D. Appleton, 1847), Supplement, 130.

2. John Collis Nesbit, *History and Properties of the Different Varieties of Natural Guanos* (London: Rogerson & Tuxford, 1859), 43.

3. For an excellent discussion of the development and reception of Liebig's mineral theory, see Margaret W. Rossiter, *Emergence of Agricultural Science: Justus Liebig and the Americans, 1840–1880* (New Haven: Yale University Press, 1975),

29–46. Also see *Hunt's Merchant's Magazine* 41 (1859): 477; and "Liebig and Laws Controversary," *American Farmer* s. 4, 14 (1858–59): 213–15.

    4. United States Guano Company, *Report to the Stockholders* (New York, 1859): 35.

    5. Advertisement for Swan Islands guano, *American Agriculturist* (1859): 255; advertisement for Philadelphia Guano Company, *American Farmer* 13 (July 1857), n.p.

    6. Ure, *Dictionary of Arts, Manufacturing, and Mines* (1847), "Supplement," 131, quoted in American Guano Company, *Report to the Stockholders of the American Guano Company* (Brooklyn: Jacobs & Brockwag, 1857), 16.

    7. U.S. Gauno Company, *Report*, 38–39.

    8. Wood & Grant, *Phosphatic Guano, from Sombrero Island* (New York, [1857?]), cover, 11-16.

    9. March advertisements, *American Farmer* s. 4, 13 (1857–58): n. p.

    10. William H. Webb, *Guano from Baker's and Jarvis Islands in the Pacific Ocean Imported by William H. Webb* (New York: Slote & James, 1862), 3–4, 20–23.

    11. U.S. Guano Company, *Report*, 39, 16, 38–39.

    12. For example, see John B. Sardy's advertisement for William Webb in *American Agriculturist* 18 (1859): 378.

    13. Eugene Gaussoin, *Memoir on the Island of Navassa (West Indies)* (Baltimore: N. pub., 1866). 26.

    14. *American Agriculturist* 18 (1859): 199, 294.

    15. Ibid., 375.

    16. Ibid., 19 (1860): 133.

    17. Ibid., 68.

    18. American Institute, *Transactions* (1862): 398.

    19. For examples of variable results in the field and laboratory, see *American Farmer* s. 4, 12 (1856–57): 379; 13 (1857–58): 53; American Guano Company, *Report to the Stockholders*, 39; Samuel Johnson, *Essays on Peat, Muck, and Commercial Manures* (Hartford: Brown & Gross, 1859). 20. Some of the chemists employed by the guano companies may have altered or misrepresented their analyses to suit the interests of the companies. For example, Samuel Johnson once charged that an analysis by a chemist later employed by the American Guano Company was "eminently adapted to deceive" (Johnson, *Essays*, 19).

    Still another possible problem was that some advertisements simply stretched the truth. For example, the American Guano Company's standard advertisement in 1860 stated that its guano had "ammonia sufficient to produce immediate abundant crops" even though the company's own analysis showed that their guano contained only traces of ammonia. (For examples, see *American Agriculturist* 19 [1860]: 37, 127.) Similarly, 1860 advertisements for Swan Island guano incorrectly noted its "richness in organic matter" (*American Agriculturist* 19 [1860]: 30). Some advertisements made even more startling and unsupportable assertions that these guanos would help retain moisture in the soil or that they would cause plants to be "free of insects." *American Agriculturist* 18 (1859): 378; 19 (1860): 30.

    20. Johnson, *Essays*, 177.

    21. For examples, see U.S. Patent Office, *Agricultural Report* (1857), 101; *American Farmer* s. 4, (1853–54): 275; February and March advertisements, *American Farmer*, 12 (1856–57): 290.

22. John Kettlewell and John S. Reese began manipulating guano in 1853 or 1854. See frequent advertisements for DeBurg, Kettlewell, and Reese, *American Farmer*, s. 4, 12 (1856–57), 13 (1857–58), and 14 (1858–59). Also see advertisements at the end of *Southern Planter* 20 (1860), passim.

23. *Southern Planter* 20 (1860): 506; *Southern Cultivator* 18 (1860): 81; *American Agriculturist* 19 (1860): 159.

24. See advertisement of Edmond Davenport & Company of Richmond in *Southern Planter* 20 (1860): advertising section, 7; and DeBurg's advertisement, *American Farmer* s. 4, 12 (1856–57): cover.

25. Several of the manipulators advertised that their product contained 8 percent ammonia and 50 percent phosphate of lime. Since the phosphate of lime was largely insoluble, however, the available phosphoric acid rating would have been much lower. See *Southern Cultivator* 18 (1860: 505; and Reese's May advertisement, *American Farmer* s. 4, 12 (1856–57): n.p.

26. *Southern Planter* 20 (1860): 505–506.

27. For examples of claims and counterclaims about the solubility of the phosphatic guanos see U.S. Patent Office, *Agricultural Report* (1854), 95; Philadelphia Guano Company advertisement, *American Farmer* s. 4, 13 (July 1857); *American Farmer* 13 (1857–58): 79–81; Philadelphia Guano Company, *Colombian Guano, Brought from the Guano Islands in the Caribbean Sea Belonging to the Republic of Venezuela* (Philadelphia: James H. Bryson, 1856), 6–7; Johnson, *Essays*, 37.

28. See chapter 7 for more detailed treatment of this subject and for references.

29. *National Cyclopaedia of American Biography*, 25: 63; advertisement, *American Farmer* s. 4, 9 (1853–54): 304; March advertisements, *American Farmer* 12 (1856–57): n.p.; John Thomas Scharf, *History of Baltimore City and County* (Philadelphia: Louis H. Everts, 1881), 399–400; George W. Howard, *The Monumental City, Its Past History and Present Resources* (Baltimore: J.D. Ehlers, 1873), 240; *Maryland Directory* (Baltimore: J. Frank Lewis, 1882), 228.

30. Scharf, *History of Baltimore*, 398; advertisement, *Maryland Directory* (1882); Kelley, *Federal Hill Story* notes, Maryland Historical Society Library, "Rasin," 3–4.

31. *American Farmer* 9 (1853–54): 295.

32. In 1868, he listed himself as an agent for the sale of "Orchilla guano" from Venezuela. (*Baltimore City Directory* [1868–69], 708; Howard, *Monumental City*, 236.)

33. Johnson, *Essays*, 27, 32.

34. Scharf, *History of Baltimore*, 400; Kelley, *Federal Hill Story* notes, "Ober," 1.

35. Scharf, *History of Baltimore*, 398–99.

36. *American Farmer* s. 4, 12 (1856–57): 322 and March advertisements, n. p.; September advertisements, *American Farmer* 13 (1857–58): n. p.; John S. Reese & Company, Broadside (Gettysburg, Pa., 1866); *Maryland Directory*, vi; *Matchett's Baltimore Directory, 1849–50*, s.v.

37. Pacific Guano Company, *The Pacific Guano Company, Its History: Its Products and Trade: Its Relation to Agriculture* (Cambridge: Riverside Press, 1876), 7–8; G. Brown Goode, "A History of the Menhaden," U.S. Commission of Fish and Fisheries, *Report of the Commissioner for 1877* (Washington: GPO, 1879), 487–88; Reese & Company, broadside. Reese, who became the Pacific Guano Company's agent for the southern states in 1864 or 1865, was still apparently advertising an

unprocessed guano in 1866. The New York agent, J. O. Baker, however, was already advertising "Ammoniated Pacific Guano" in May 1865. See his advertisement, *American Agriculturist* 24 (1865): 97.

38. Pacific Guano Company, *Pacific Guano Company*, 9.

39. Goode, "History of the Menhaden," 487.

40. Pacific Guano Company, *Pacific Guano Company*, 24, 34; J. M. Glidden to A. F. Crowell, 28 March 1877, and W. T. Glidden to J. M. Glidden, 5 October 1874, Pacific Guano Company Collection. The company's promotional pamphlet issued for the 1876 Centennial Exhibition in Philadelphia (Pacific Guano Company, *Pacific Guano Company*, 25–30) contains an elaborate description of the Swan Islands and indicates that the company was still trying to find some profitable use for the islands, possibly as a free port, coaling station, or marine hospital.

41. Goode, "History of the Menhaden," 488. The company records show an occasional cargo of Navassa guano was purchased. For example, see Pacific Guano Company, "Daybook, 1873–1877," Manuscript, Pacific Guano Company Collection, Baker Library, Harvard Business School, 44.

42. For example, see Gaussoin, *Memoir* 24; and Navassa Phosphate Company, *Navassa Phosphate Company* (Baltimore: N. pub., 1864), 7, 9.

43. Williams Haynes, *American Chemical Industry* (New York: D. Van Nostrand Company, 1954), vol. 1, 345–46.

44. Pacific Guano Company, *Pacific Guano Company*, 34.

45. *Fertilizer Review* (July–August 1938).

Chapter VI: Animal and Fish Guano

1. *Plough, Loom, and Anvil* 3 (1850–51): 638.

2. *Cultivator*, n.s., 7 (1850): 197–98.

3. *American Farmer* s. 4, 6 (1850–51): 430; *Cultivator* s. 3, 4 (1856): 220; Maryland State Agricultural Chemist, *Second Report of James Higgins, M.D., State Agricultural Chemist to the House of Delegates of Maryland* (Annapolis: Thomas E. Martin, 1852), 82.

4. *Country Gentleman* 7 (1856): 362; *American Farmer* s. 4, 13 (1857–58): 49.

5. *Cultivator* n.s., 7 (1850): 197; November advertisements, *Cultivator* 8 (1852–53): 38.

6. *Country Gentleman* 7 (1856): 282, 362. British manufacture began about 1850 (Rowland E. Prothero Ernle, *English Farming Past and Present* (London: Longmans, Green, 1917), 371.

7. *American Farmer* s. 4, 12 (1856–57): 383–84.

8. American Institute of the City of New York, *Transactions* (1860): 462; *American Farmer* s. 4, 13 (1857–58): 18.

9. *American Agriculturist* 17 (1858): 95.

10. Ibid. 19 (1860): 30

11. Ibid. 16 (1857): 71; Henry R. Stiles, *The Civil, Political, Professional, and Ecclesiastical History, and Commercial and Industrial Record of the County of Kings and the City of Brooklyn from 1683 to 1884*, 2 vols. (New York, 1884), 78.

12. Stiles, *History of Brooklyn*, 78, 756.

13. *American Agriculturist* 16 (1857): 71.

14. Ibid. 18 (1859): 127.

15. Ibid. 24 (1865): 30; 25 (1866): 159, 160.

16. Ibid. 25 (1866): 162.

17. See John Duffy, *A History of Public Health in New York City, 1866–1966* (New York: Russell Sage Foundation, 1974): 24–25, 128–29, for a good account of the problems posed by slaughterhouses in New York City.

18. Henry I. Bowditch, *Public Hygiene in America* (Boston: Little Brown, 1877), 50–51; Brighton Abattoir, *A Description of the Brighton Abattoir and the Animal Fertilizer Made by the Butchers' Slaughtering & Melting Association of Brighton, Mass.* (Boston: J. A. Cummings, 1874), 1.

19. Abattoir, *Description*, 3. See also MSBA, *Annual Report* 1 (1873): 360.

20. Williams, Clark & Company *Williams, Clark, & Co., Manufacturers of High Grade Bone Fertilizers* (New York: William J. Schaufels, 1886), n. p.

21. Swift & White, *Swift & White, Manufacturers of and Dealers in Fertilizers of All Descriptions* (New York: Robert Malcolm, 1876), passim; and *American Agriculturist* 34 (1875): 119.

22. Williams, Clark, & Company, *Williams, Clark & Co.*, n. p.

23. Williams Haynes, *American Chemical Industry* (New York: D. Van Nostrand Company, 1954), 1: 351n; R. A. Clemen, *The American Livestock and Meat Industry* (New York: The Ronald Press, 1923), 367–68.

24. *American Farmer* s. 4, 13 (1857–58): 49.

25. Stiles, *History of Brooklyn*, 757.

26. NYSAS, *Transactions* 2 (1842): 208.

27. Charles T. Jackson, U.S. Patent Office, *Agricultural Report* (1844), 380.

28. Ibid.

29. *Mark Land Express*, reprinted in *American Agriculturist* 11 (1853–54): 347–48, 357.

30. NYSAS, *Transactions* (1849): 671.

31. Jackson, U.S. Patent Office, *Agricultural Report* (1844): 380.

32. Roger W. Harrison, *The Menhaden Industry*, U.S. Department of Commerce, Bureau of Fisheries (Washington: GPO, 1931), 6.

33. G. Brown Goode, "A History of the Menhaden," U.S. Commission of Fish and Fisheries, *Report of the Commissioner for 1877* (Washington: GPO, 1879), 162, 374, 376, 603; G. Brown Goode and A. Howard Clark, "The Menhaden Fishery," *The Fisheries and Fishery Industries of the United States* (Washington: GPO, 1887): 365.

34. Goode and Clark, *"Menhaden Fishery,"* 366.

35. According to the AMOGMM, *The Menhaden Fishery of Maine* (Portland: B. Thurston, 1878), 17, the earliest operators threw the fish scrap back into the water.

36. For speculation on the role of the whale oil scarcity, see *American Agriculturist* 26 (1867): 400.

37. *Cultivator* s. 3, 4 (1856): 259; W. O. Atwater, "Menhaden and Other Fisheries and Their Products as Related to Agriculture," U.S. Commission of Fish and Fisheries, *Report for the Commissioner for 1877* (Washington: GPO, 1879); Goode, "History of Menhaden," 108; *Country Gentleman* 8 (1856): 43; James F. W. Johnston, *Notes on North America: Agricultural, Economical, and Social* (Edinburgh: Blackwood, 1851), 2, 232.

38. Goode, "History of Menhaden," 444; Charles R. Stark, *Groton, Connecticut: 1705–1905* (Stonington, Conn.: Palmer Press, 1922), 396–97.

39. Goode, "History of Menhaden," 163, 366, 446, 448; Gilbert Burling,

"Long Island Oil Fishery," unidentified clipping (2 September 1871), East Hampton Free Library, 268; *Republican Watchman* (Greenport, N.Y.), 11 December 1869.

40. *Cultivator* n.s., 8 (1851): 218.

41. Goode, "History of Menhaden," 163–64, 494; Goode and Clarke, "Menhaden Fishery," 369.

42. U.S. Patent Office, *Agricultural Report* (1854): 107; *Country Gentleman* 8 (1856): 43; *American Agriculturist* 11 (1853–54): 292–93; 12 (1854): 52; *Working Farmer* 6 (1854): 70; 7 (1856): 236; American Institute *Transactions* (1855): 345; *Cultivator* s. 3, 3 (1855): 333; *Southern Planter* 15 (1855): 169.

43. *Country Gentleman* 8 (1856): 49; Atwater, "Menhaden," 209.

44. Ibid., 209, 212.

45. Ibid., 212; *Country Gentleman* 8 (1856): 43; *American Agriculturist* 11 (1853–54): 292–93.

46. Atwater, "Menhaden," 226.

47. *American Agriculturist* 12 (1854): 52, lists several other English processes.

48. U.S. Patent Office, *Agricultural Report*, 107; *De Bow's* 19 (1855): 235; *Maine Farmer*, quoted in *Southern Planter* 15 (1855): 169. Either a second company began operations in Bristol, R.I., about the same time, or a branch of Halliday's company operated there. See *Cultivator* s. 3, 3 (1855): 351.

49. *Plough, Loom, and Anvil* 9 (1856–57): 39.

50. Advertisement, *American Agriculturist* 15 (1856): 181.

51. S. W. Johnson, "On Fish Manures," *Country Gentleman* 8 (1856): 43–44.

52. Johnson, "On Fish Manures," 43–44.

53. Advertisement, *American Agriculturist* 15 (1856): 181.

54. Johnson, "On Fish Manures," 43–44.

55. Atwater, "Menhaden," 210.

56. Johnson, "On Fish Manures," 43–44.

57. *American Agriculturist* 17 (1858): 68, 95; Long Island Fish Guano & Oil Works, *Patent Fish Guano* (1858), n.p.; *American Institute* (1857): 564.

58. Long Island Fish Guano, *Patent Fish Guano*, n.p.

59. Ibid.

60. *American Agriculturist* 17 (1858): 68.

61. Goode, "History of Menhaden," 162–69.

62. See the interesting diary, 1852–1880, of B. F. Conklin, Jamesport, Long Island, in Goode, "Menhaden Fishery," 372–415, for an example of how a farmer-fisherman gradually specialized in the supply of fish for oil factories in the area.

63. William H. Glover, Diary, 1877–1878, East Hampton Free Library.

64. Goode and Clark, "Menhaden Fishery," 368; Goode, "History of Menhaden," 163, 192, 446.

65. Stiles, *History of Brooklyn*, 78. Frank Swift occupied the rendering and animal fertilizer plant established earlier by William B. Reynolds (Stiles, *History of Brooklyn*, 78). In 1877, Swift was listed in Goode, "History of Menhaden," 168, as the operator of a fish guano plant on the island. He may later have been a partner in Swift & White, the integrated fertilizer producer on the island. There may have been other similar connections between animal fertilizer plants, fish guano plants, and phosphate plants.

66. Goode, "History of Menhaden," 164; Atwater, "Menhaden," 210–11; Francis Byron Green, *History of Boothbay, Southport, and Boothbay Harbor, Maine, 1623–1905* (Portland: Loring, Short & Harmon, 1906), 371; AMOGMM, *Menhaden Fishery of Maine*, 16.

67. Goode and Clark, "Menhaden Fishery," 369. Northern operators made most of the first attempts to establish the industry in the South (Harrison, *Menhaden Industry*, 10).

68. *American Agriculturist* 26 (1867): 400.

69. Ibid.

70. *Republican Watchman*, 11 December 1869.

71. For descriptions of fish factories in Long Island and Maine, see Burling, "Long Island Oil Fishery," 268–69; *American Agriculturist* 26 (1867): 400; 27 (1868): 451–52; Goode and Clark, "Menhaden Fishery," 344; AMOGMM, *Menhaden Fishery of Maine*, 10; *Sag Harbor Corrector*, 12 July 1879, n.p.; Goode, "History of Menhaden," 170–74.

72. Goode, "History of Menhaden," 176.

73. Ibid., 174; Green, *History of Boothbay*, 372–73.

74. Burling, "Long Island Oil Fisheries," 269. Burling, who had been assigned to write an article on the industry, was clearly relieved when his plant tour was over and he could go out on the fishing boats to cover the more glamorous (and less odoriferous) part of the industry.

75. *Glen Cove Gazette*, 14 Aug. 1869.

76. Goode, "History of Menhaden," 490.

77. *Roslyn News*, 19 August 1882.

78. Harrison, *Menhaden Industry*, 6.

79. *Sag Harbor Corrector*, 12 July 1879, n.p.

80. Burling, "Long Island Oil Fishery," 269; Goode, "History of Menhaden," 191–92.

81. AMOGMM, *Menhaden Fishery of Maine*, 16.

82. *Sag Harbor Corrector*, 12 July 1879, n.p.

83. B. F. Conklin, Diary, in Goode and Clark, "Menhaden Fishery," 372; AMOGMM, *Menhaden Fishery of Maine*, 16. For descriptions of fishing operations, see *American Agriculturist* 27 (1868): 451–52; Burling, "Long Island Oil Fishery," 270–73; Goode, "History of Menhaden," 176.

84. Green, *History of Boothbay*, 272–73.

85. *American Agriculturist* 34 (1875): 250. See survey of fish guano use in Charles W. Smiley, "The Extent and Use of Fish Guano As a Fertilizer," *Report of the Commissioner of Fish and Fisheries for 1881*, U.S.Congress, 47: 1, *Senate Miscellaneous Document* No. 10 (1884), 663–95. A c. 1867 advertising broadside of the Old Rockaway Oil and Guano Works, located on Barren Island, suggested that farmers from around Jamaica Bay should take their supplies of fish guano directly from the factory.

86. *American Agriculturist* 26 (1867): 169.

87. Noah Youngs, Day Book, 1847, manuscript in private collection, Riverhead, N.Y.

88. Some farmers apparently even thought the strong odor drove off injurious insects. (Gilbeart H. Collings, *Commercial Fertilizers: Their Sources and Use* [Philadelphia: P. Blakiston's Son, 1938], 125.)

89. *Working Farmer* 7 (1855): 180; *American Agriculturist* 11 (1853–54): 347–48, 357.

90. See S. W. Johnson, *Cultivator* s. 3, 4 (1856): 259; *Working Farmer* 7 (1855): 180; AMOGMM, *Menhaden Fishery of Maine*, 28–30; *American Agriculturist* 34 (1875): 141; Pacific Guano Company, *The Pacific Guano Company: Its History, Its Products, and Trade: Its Relation to Agriculture* (Cambridge: Riverside Press, 1876), 9.

91. AMOGMM, *Menhaden Fishery of Maine*, 30.

92. *Cultivator* s. 3, 4 (1856): 259.

93. AMOGMM, *Menhaden Fishery of Maine*, 34.

94. Goode, "History of Menhaden," 424, 434.

95. *Republican Watchman*, 11 December 1869, n.p.; Burling, "Long Island Oil Fishery," 269.

96. See advertisement for "prepared fish guano" in *American Agriculturist* 34 (1875): 112.

97. Goode, "History of Menhaden," 487.

98. Pacific Guano Company, Letterbook No. 8, 1869–74, manuscript in Baker Library, Harvard Business School, contains numerous letters pertaining to fishing operations the Crowell family held interests in.

99. Goode, "History of Menhaden," 488; Pacific Guano Company, "Daybook, 1873–1877," manuscript in Baker Library, Harvard University Business School. Contains frequent references to fish scrap purchases from guano companies.

100. Harold B. Clifford, *The Boothbay Region, 1906–1960* (Freeport, Maine: Bond Weelwright, 1961), 49; *Lewiston Evening Journal*, 17 December 1874, quoted in Goode, "History of Menhaden," 491; *Country Gentleman* 29 (1867): 89.

101. Atwater, "Menhaden," 227; Goode, "History of Menhaden," 493.

102. Collings, *Commercial Fertilizers*, 125.

103. *American Agriculturist* 26 (1867): 164.

104. MSBA, *Annual Report* (1870): 196.

105. Green, *History of Boothbay*, 373.

106. *Roslyn News*, 19 August 1882, n.p.

107. Goode, "History of Menhaden," 190.

108. For a table showing the production and consumption of fishery by-products from 1850 to 1953, see A. L. Mehring, J. Richard Adams, and K. D. Jacob, *Statistics on Fertilizers and Liming Materials in the United States*, USDA, Agricultural Research Service, Statistical Bulletin No. 191 (April 1957), 36.

Chapter VII: Superphosphates

1. Basically, the treatment with the sulfuric acid converts the insoluble tricalcium phosphate (bone phosphate of lime) into the more soluble monocalcium phosphate. Gypsum is a major by-product and constitutes about half of the weight of a superphosphate. The actual chemical reactions are complex, with a number of intermediary products and two-way reactions involved. For material on the beginnings of superphosphate manufacture in Europe see William H. Waggaman, *Phosphoric Acid, Phosphates, and Phosphatic Fertilizers* (New York: Reinhold, 1952), 258; and Gilbeart H. Collings, *Commercial Fertilizers: Their Sources and Use*

(Philadelphia: P. Blakiston's Son, 1938), 3; "History and Technique of the Super-phosphate Industry," *Superphosphate* 1 (1928): 94–95; Max Speter, "Final Sum-mary of the Research into the Origin of Superphosphates," *Superphosphate* 8 (1935): 142–43; K. D. Jacob, "History and Status of the Superphosphate Indus-try," USDA and TVA, *Superphosphate: Its History, Chemistry, and Manufacture* (Washington: GPO, 1964), 19–20.

2. Justus Liebig, *Organic Chemistry in Its Applications to Agriculture and Physiology*, 2nd American ed. (Cambridge: John Owen, 1841), 175.

3. Speter, "Origin of Superphosphates," 162. Liebig's fertilizer was the oppo-site of a superphosphate. He attempted to make his phosphates insoluble to prevent leaching rather than soluble to enhance absorption by plants.

4. Ibid., 183–86. For a more extensive discussion of Lawes, see Sir E. John Russell, *A History of Agricultural Science in Great Britain: 1620–1954* (London: George Allen & Unwin, 1966), 91–95.

5. *Superphosphate* 1 (1928): 99.

6. Ibid., 112.

7. Jacob, "History and Status," 30.

8. By 1863, British farmers were purchasing nearly six million dollars worth of superphosphates (*American Agriculturist* 22 [1863]: 135).

9. *Cultivator* n.s., 1 (1844): 364. For an account of Norton's stay in Edinburgh, see Margaret W. Rossiter, *Emergence of Agricultural Science: Justus Liebig and the Americans, 1840–1880* (New Haven: Yale University Press, 1975), 96–100.

10. *American Agriculturist* 3 (1844): 251.

11. U.S. Patent Office, *Agricultural Report* (1845): 1051; *American Farmer* s. 4, 1 (1845–46): 12, 44; 3 (1847–48): 64; *American Agriculturist* 5 (1846): 206, 228, 291; 6 (1847): 138, 355; *Cultivator* n.s., 3 (1846): 110; NYSAS, *Transactions* 7 (1847): 376–77, 380–82.

12. NYSAS, *Transactions* 9 (1849): 665–66.

13. *Cultivator* n.s., 7 (1850): 277; 8 (1851): 52–53, 270–71; s. 3, 3 (1855): 83; *Country Gentleman* 5 (1855): 69.

14. *Plough, Loom, and Anvil* 4 (1851): 119–25; NYSAS, *Transactions* 11 (1851): 304–40; *Working Farmer* 3 (1851): 123, 209, 221–22, 267.

15. *Cultivator* s. 3, 2 (1854): 130, 167; *American Farmer* s. 4, 4 (1848–49): 378; 6 (1850–51): 237, 252; November advertisements, *American Farmer* 8 (1852–53): 38; 9 (1853–54): 304.

16. *Working Farmer* 3 (1851): 72.

17. *Cultivator* n.s., 1 (1844): 172.

18. For Chappell, see *American Farmer* s. 4, 4 (1848–49): 290; 5 (1849–50): 45. For Kettlewell & Davison, see *American Farmer* s. 4, 4 (1848–49): 364, 377.

19. Jacob, "History and Status," 43; William J. Kelley, "Phillip S. Chappell," *Federal Hill Story* papers, manuscript 1692, Maryland Historical Society, 1; Kelley, "Davison, Kettlewell & Co.," *Federal Hill Story* papers, 2; *Baltimore* (April 1938): 32–33.

20. *American Farmer* s. 4, 5 (1849–50): 290; 6 (1850–51): 280, 318; 8 (1852–53): 343, November advertisements, 38–39.

21. Ibid. s. 4, 5 (1849–50): 45.

22. Ibid., 261; *Southern Planter*, quoted in *Cultivator* n.s. 8 (1851): 60.

23. *American Farmer* s. 4, 6 (1850–51): 318, 321. These advertisements seem to be aimed at dispelling adverse publicity from experimental failures.

24. Ibid., 451; 7 (1851–52): 351; 9 (1853–54): 272.

25. See Kettlewell's advertisement in every issue of *American Farmer* s. 4, 13 (1857–58). Chappell apparently stopped producing fertilizers in the mid-1850s. He continued to produce sulfuric acid, and according to R. W. L. Rasin, was the only producer in Baltimore in 1861 (Rasin, "Growth of the Fertilizer Industry," *American Fertilizer* 1 [1894], 7). William Davison also apparently temporarily left the fertilizer business. According to *Marchett's Baltimore Directory, 1855–56*, 69 and 89, Davison was a manufacturer of chemicals, varnishes, paints, etc., and Chappell was a chemist. Neither listing mentioned fertilizers.

26. William H. Shaw, *History of Essex and Hudson Counties, N.J.* (Philadelphia: Everts & Peck, 1884), 660. For biographies, see *ibid.*, 657–66; Carl R. Woodward, *The Development of Agriculture in New Jersey, 1640–1880* (New Brunswick, N.J.: Rutgers University, 1927), 127–50; *National Cyclopaedia of American Biography* (Clifton, N.J.: J. T. White, 1979), 3: 178; and Williams Haynes, *Chemical Pioneers: The Founders of the American Chemical Industry* (New York: D. Van Nostrand, 1939), 74–87. The Farmer's Club of the American Institute awarded Mapes a silver cup for the best cultivated twenty-five acre farm. The visiting committee noted "perhaps no farm in the vicinity of New York has been more improved within the last five years than this one" (American Institute of the City of New York, *Transactions* [1852]: 49–54). For a critical picture of Mapes, see Rossiter, *Emergence of Agricultural Science*, 150–55. Mapes was born on Long Island and spent part of his youth at a boarding school in Hempstead, where he resided for a year with William Cobbett, the English radical and agricultural reformer, during the latter's stay in the United States.

27. For example, see *Working Farmer* 1 (1849): 26; 2 (1850): 43.

28. Ibid. 11 (1859): 80, 97. See Johnson, *Essays on Peat, Muck, and Commercial Manure* (Hartford: Brown & Gross 1859), 178; and *Country Gentlemen* 20 (1862): 361, for Johnson's criticism of this theory.

29. *Working Farmer* 2 (1850): 5, 191.

30. Ibid. 1 (1849): 26; 2 (1850): 43.

31. Ibid. 3 (1851): 124

32. Ibid. 1 (1849): 56; 2 (1850): 43.

33. Ibid. 3 (1851): 75.

34. For example, see the lecture by the British chemist James Thomas Way before the NYSAS, reported in *Working Farmer* 3 (1851): 123. Lawes had used bone charcoal as the raw material in some of his first superphosphates. See Russell, *History of Agricultural Science* 94.

35. *Working Farmer* 1 (1849) 26; *National Cyclopaedia of American Biography*, 3: 178; Jacob, "History and Status," 38.

36. U.S. Department of Interior, Bureau of the Census, *Manufactures of the United States in 1860* (Washington: GPO, 1865), counted eighteen sugar refineries in greater New York, although it found none in Newark that year.

37. *Working Farmer* 2 (1850): 43–44.

38. Waggaman, *Phosphoric Acid*, 240.

39. Gilbeart H. Collings, *Commercial Fertilizers*, 170.

40. *Working Farmer* 3 (1851): 75.

41. Ibid. 1 (1849): 87.
42. Ibid. 4 (1852): 27.
43. Ibid. 3 (1851): 221.
44. Ibid. 4 (1852): 27.
45. Ibid.
46. Ibid., 72.
47. Ibid., 115, 117, 118, 136, 142, 165, 178, 195, 202, 130.
48. Ibid., 27, 281.
49. Ibid. 11 (1859–60): 49.
50. Potential analysis calculated by using the maximum possible analyses of ingredients and assuming no addition of water or conditioners.
51. Johnson, 28.
52. *Country Gentleman* 1 (1853): 131.
53. Johnson, *Essays*, 28.
54. Early phases of this dispute are summarized in *Cultivator* s. 3, 2 (1853): 149–50. See also *Working Farmer* 6 (1854): 3–4, for related disputes between Joseph Harris and J. J. Mapes. The *Country Gentleman* 20 (1862): 315, relates the entire history of the dispute.
55. *Working Farmer* 6 (1854): 3. See also *Genesee Farmer*, quoted in *Working Farmer* 6 (1854): 166.
56. For articles on the Chilean guano fraud, see *Country Gentleman* 5 (1855): 117–18, 296, 343–44, 360, 405; 6 (1855): 25, 76, 93, 122; *Working Farmer* 7 (1855): 126–27, 169; *American Farmer* s. 4, 11 (1855–56): 22, 49. See discussion of this in chapter 5.
57. For example, see *Massachusetts Ploughman*, quoted in *Working Farmer* 4 (1854): 70; *Country Gentleman* 14 (1859): 283, 348, 389; Johnson, *Essays*, 82; *American Agriculturist* 21 (1862): 139. In 1859, Johnson called Mapes's superphosphate a "series of trashy mixtures" (quoted in *Country Gentleman* 23 [1864]: 266). In 1863, Evan Pugh of the Pennsylvania Agricultural College criticized Mapes for "gross dishonesty" (*Country Gentleman* 21 [1863]: 298).
58. For example, see *Mapes' Nitrogenized Super-Phosphate of Lime* (1859), which cites thirty advantages of his fertilizer over barnyard manure, including that it prevented animals from suffering bone disease, that insects disliked it, and that it had "hygrometric power."
59. *Working Farmer* 3 (1851): 75.
60. In addition to manufacturing the "Chilean guano," he had been involved in what one Waterville, New York, farmer called the "humbug" of "Stowell Evergreen Corn" (*Country Gentleman* 1 [1853]: 180).
61. See *Pennsylvania Farm Journal*, quoted in *Working Farmer* 6 (1854): 126; *Working Farmer* 6 (1854): 32.
62. *Working Farmer* 5 (1853): 73; 6 (1854); 32, 199.
63. *American Agriculturist* 11 (1853): 208; Eugene L. Armbruster, *Brooklyn's Eastern District* (Brooklyn, 1842), 211; Henry R. Stiles, *The Civil, Political, Professional, and Ecclesiastical History, and Commercial and Industrial Record of the County of Kings and the City of Brooklyn from 1683 to 1884*, 2 vols. (New York, 1884), 756.
64. Stiles, *History of Brooklyn*, 756.
65. *Country Gentleman* 1 (1853): 111; Haynes, *American Chemical Industry*, 1: 347.

66. Armbruster, *Brooklyn's Eastern District*, 211; Stiles, *History of Brooklyn*, 756.

67. Armbruster, *Brooklyn's Eastern District*, 210.

68. *American Agriculturist* 12 (1854): 56; 14 (1855): 62.

69. In 1844, the British chemist, James F. W. Johnston, described a method of treating either fermented urine or gashouse waste ammoniacal liquor with sulfuric acid to produce ammonium sulfate (*Lectures on the Applications of Chemistry and Geology to Agriculture* [New York: Wiley & Putnam, 1844], 350). See also John Hannam, *The Economy of Waste Manures* (London: Longman, Brown, Green and Longmans, 1844), 86. The utility of gashouse wastes as a fertilizer had been known for much longer. See *American Farmer* 10 (1828–29): 9; n.s., 4 (1837–38): 370.

70. The editors of the *American Agriculturist*, after diligent inquiry, were able to locate only one manufactory of ammonium sulfate in 1854. That one was in Williamsburg, apparently De Burg's (*American Agriculturist* 12 [1854]: 56).

71. For example, see Longett I. Griffing advertisement, *Country Gentleman* 1 (1853): 351.

72. *Southern Planter*, quoted in *American Farmer* s. 4, 12 (1856–57): 391.

73. Johnson, *Essays*, 29.

74. *American Farmer* s. 4, 13 (1857–58): 29.

75. Ibid.

76. Ibid.

77. Ibid., 49.

78. Armbruster, *Brooklyn's Eastern District*, 211; Stiles, *History of Brooklyn*, 757. The Coe family apparently had numerous connections in the fertilizer business. Andrew Coe of Middleton, Connecticut, began manufacturing fertilizer about 1854. Russel Coe and Elmer F. Coe founded Coe & Company in Boston in 1857. It was later taken over by William Bradley, who was married to a daughter of Calvin Coe of Meriden, Connecticut. An Andrew Coe, supposedly of Boston, worked for Armor Smith in Cincinnati. E. Frank Coe took over De Burg's Brooklyn plant in 1860 and used Enoch Coe as his agent. C. C. Coe was supervisor of Bradley's South Carolina plant in the 1870s while Russell Coe produced a highly reputed superphosphate in New Jersey. See Jacob, "History and Status," 42; *National Cyclopaedia of American Biography*, 24: 236; *American Agriculturist* 28 (1869): 153; *Cultivator* s. 3, 2 (1854): 167; *Country Gentleman* 29 (1867): 89.

79. Even in 1870, the 126 fertilizer manufacturing establishments listed in the U.S. Census had only 69 steam engines and 33 waterwheels. (U.S. Census Office, *Ninth Census*, vol. 3, *Statistics on Wealth and Industry of the United States* [Washington: GPO, 1872], 395).

80. See *American Agriculturist* 12 (1854): 56; R. W. L. Rasin, "Growth of the Fertilizer Industry," *American Fertilizer* 1 (1894): 7.

81. *Working Farmer* 11 (1859–60): 98.

82. For Patterson, see *American Agriculturist* 12 (1854): 75. For Andrew Coe, see *Cultivator* s. 3, 2 (1854): 167. For New York Superphosphate Manufacturing Company, see *Cultivator* and *American Farmer* 9 (1853–54): 296. For Eagle Chemical Works, see *American Farmer*, 304, and *Southern Cultivator* 11 (1853): 374.

83. February advertisement for Rogers, *American Farmer* 9 (1854): n.p. *National Cyclopaedia of American Biography*, 19: 101 for Baugh; Haynes, *American Chemical Industry*, 1: 343 and Edwin T. Freedley, *Philadelphia and Its Manufacturers*

in 1867 (Philadelphia: Edward Young, 1867), 286 for Crosdale. According to Freedley, *Philadelphia and Its Manufacturers in 1857* (Philadelphia: Edward Young, 1859), 145, there were seven plants manufacturing superphosphates in Philadelphia in 1857, but he only names two and I have only been able to identify two others.

84. See Johnson, *Essays*, 31, 33, 168, for Buck, Coe & Company, L'Loyd, Pike & Company, and Green & Preston; Jacob, "History and Status," 44, for Hatch; Haynes, *American Chemical Industry*, 1: 343, for Potts & Klett; April advertisements, *American Farmer* s. 4., 12 (1856–57): n. p., for Hildreth; ibid., 13 (1857–58) and *American Agriculturist* 18 (1859): 127, for New Jersey Superphosphate; *American Farmer* s. 4, 13 (1857–58): 39 and *Allen & Needles Improved Fertilizer* (Philadelphia, 1866), for Allen & Needles; *Hathaway's Fertilizers or Superphosphate of Lime* 1 (Boston, 1857); *Southern Cultivator* 14 (1856): 69, for New Orleans; January advertisements, *American Farmer* s. 4, 14 (1858–59): n.p., for Tasker & Clark and for Horner; *American Farmer* s. 4, 12 (1856–57): 331, for Trego and for Rhodes; Freedley, *Philadelphia Manufacturers, 1867*, 146, for Phillips; Rasin, "Growth of the Fertilizer Industry," for a suggestion that Potts & Klett made Rhodes's superphosphate in Philadelphia; Armbruster, *Brooklyn's Eastern District*, 211, and Stiles, *History of Brooklyn*, 757, for E. F. Coe; *American Agriculturist* 19 (1860): 287 for Lister; *National Cyclopaedia of American Biography*, 24: 236 for Bradley; *Plough, Loom, and Anvil* 10 (1857–58): 3, for American Fertilizing Company; advertising sheet, *Southern Planter* 20 (1860), 4, 11, for Fowle & Company and Richmond Fertilizer Manufacturing; *Southern Cultivator* 18 (1860): 356 and Kelley, *Federal Hill Story* papers, "Davison & Kettlewell," 9 for Whitlock.

85. *National Cyclopaedia of American Biography*, 3: 178.

86. B. M. Rhodes advertisement, *American Farmer* s. 4, 12 (1856–57): 331.

87. *American Fertilizer* 1 (1894): 88; *National Cyclopaedia of American Biography*, 19: 101; Edward Butler, Jr., to K. D. Jacob, 30 December 1929, Tennessee Valley Authority Technical Library.

88. *National Cyclopaedia of American Biography*, 5: 135; Newark Board of Trade, *Newark, N.J. Illustrated, A Souvenir of the City and its Numerous Industries* (Wm. A. Baker: Newark, 1893), 272; William Ford, *The Industrial Interests of Newark, New Jersey* (New York: Van Arsdale, 1874): 213.

89. *American Agriculturist* 19 (1860): 287.

90. Newark Board of Trade, *Newark*, 272; Ford, *Newark*, 213.

91. *American Farmer* s. 4, 5 (1849–50): 289.

92. January advertisements, *American Farmer* s. 4, 14 (1858–59): n.p.

93. Pacific Guano Company, *The Pacific Guano Company: Its History, Its Products and Trade, Its Relation to Agriculture* (Cambridge: Riverside Press, 1876), 7; Stephen L. Goodale, *Commercial Manures; A Lecture Delivered before the Farmers' Convention Held in Augusta, January, 1869*, for the Cumberland Bone Company; Kelley, *Federal Hill Story* papers, "Ober & Kettlewell," 2, for Ober, "Davison & Kettlewell," 9, for Chesapeake Guano Company, Davison, Symington, & Company, and Maryland Manufacturing Company; Kelly, *Federal Hill Story* "Liebig and Gibbons," 1, for Patapsco Guano Company; *American Fertilizer* 5 (1896): 214, for Patapsco Guano Company; Jacob, "History and Status," 44 and 46, for Hubbard, Currie, and Smith; Rasin, "Growth of the Fertilizer Industry," 6, for the Richmond plant; Henry Bower, *Bower's Complete Manure* (Philadelphia: Chandler,

1867), 1; Rumford Chemical Works, *Wilson's Ammoniated Superphosphate of Lime* (1870), title page.

94. Rasin, "Growth of the Fertilizer Industry," 6.

95. For example, George Grafflin first became interested in Navassa Island, and then helped organize the Patapsco Guano Company, *American Fertilizer* 5 (1896): 214.

96. Pacific Guano Company, *Pacific Guano Company*, 7.

97. Kelley, *Federal Hill Story* papers, "Davison & Kettlewell," 9 and "Ober & Kettlewell," 1.

98. Grafflin obituary, *American Fertilizer* 5 (1896): 214; Kelley, *Federal Hill Story* papers, "Chesapeake Guano Works," 1 and "Liebig and Gibbons," 1.

99. Goodale, *Commercial Manures*, 4.

100. Bower, *Bower's Complete Manure*, 1; *National Cyclopaedia of American Biography*, 16: 216.

101. See Rumford, *Wilson's Ammoniated*, 22–23.

102. Rasin, "Growth of the Fertilizer Industry," 7.

103. Baugh & Sons, *How to Maintain the Fertility of American Farms and Plantations* (Philadelphia: Spangler & Davis, 1866), 18–19.

104. *American Farmer* s. 4, 12 (1856–57): March cover.

105. C. B. Rogers advertisement, *American Farmer* s. 4, 9 (1853–54): n. p.

106. Baugh & Sons advertisement, Freedley, *Philadelphia Manufacturers*, 1867, 288.

107. U.S. Patent Office, *Agricultural Report* (1853): 79; *Country Gentleman* 3 (1854): 150; 4 (1854): 247; 6 (1855): 239; *American Farmer* 14 (1855): 146; MSBA, *Annual Report* 2 (1857): 202.

108. For successful experiments, see NYSAS, *Transactions* 13 (1853): 584; U.S. Patent Office, *Agricultural Report* (1853): 88; *Country Gentleman* 3 (1854): 150; *American Agriculturist* 12 (1854): 89. For unsuccessful experiments, see: *Country Gentleman* 3 (1854): 55; 4 (1854): 247; *American Agriculturist* 14 (1855): 146.

109. For Mapes, see testimonials in *Working Farmer* 4 (1852): 117, 142, 165, 272; *Working Farmer* 6 (1854): 32; Mapes, *Mapes' Nitrogenized*, n. p.

110. *American Fertilizer* 1 (1894): 88.

111. *Working Farmer* 7 (1856): 280; see testimonials from Georgia and South Carolina in Mapes, *Mapes' Nitrogenized*, n. p.

112. *American Farmer* s. 4, 12 (1856–57): 49. See large advertisements in every issue of *American Farmer* s. 4, 12 (1856–57).

113. For example, a formula penciled inside the cover of Harvard's copy of *Hathaway's Fertilizer's* indicates that it consisted of 15 percent bone black treated with half its weight in sulfuric acid. This would have produced about 1.3 percent available phosphoric acid in the finished product.

114. The average of the five brands was 2.5 percent ammonia and 5.1 percent available phosphoric acid. Johnson did not bother to look for potassium, but would have found little. See Johnson, *Essays*, 28–32. Modern superphosphates contain 18–20 percent available phosphoric acid.

115. Johnson, *Essays*, 21.

116. *Country Gentleman* 4 (1854): 71.

117. *American Institute* (1858): 343.

118. Johnson, *Essays*, 27–33.

119. *American Agriculturist* 21 (1862): 139.
120. *Country Gentleman* 19 (1862): 384.
121. *American Agriculturist* 21 (1862): 139.
122. Johnson, *Essays*, 25.
123. *American Agriculturist* 16 (1857): 5–6, 53, 71; 17 (1858): 165, 328.
124. Ibid. 18 (1859): 294; 21 (1862): 139.
125. Ibid. 22 (1863): 168.
126. Ibid. 27 (1868): 220.
127. Ibid. 22 (1863): 168.
128. Johnson, *Essays*, passim.
129. NYSAS, *Transactions* 20 (1860): 238; *Country Gentleman* 10 (1857): 10; 14 (1859): 299; MSBA, *Annual Report* 2 (1858): 190; 2 (1857): 202; 1 (1868): 285; *Southern Cultivator* 18 (1869): 356; *American Agriculturist* (1860): 107.
130. See *American Agriculturist* (1854): 89; Freedley, *Philadelphia Manufacturers 1867*, 285, estimated that two million dollars worth of fertilizers were manufactured annually in Philadelphia.

Chapter VIII: From Chaos to Stability

1. Charles A. Groessman, address to the NYSAS, 22 January 1873, reprinted in *Country Gentleman* 38 (1873), 83; S. L. Goodale, *Commercial Manures, a Lecture Delivered Before the Farmers' Convention Held at Augusta, January, 1869* [1869], 14.

2. For discussions of the role of phosphorus as the first limiting nutrient on most soils, see Mirko Lamer, *The World Fertilizer Economy* (Stanford, Calif.: Stanford University Press, 1957), 40 and Gilbeart H. Collings, *Commercial Fertilizers: Their Sources and Use* (Philadelphia: P. Blakiston's Son, 1938), 202.

3. "History and Technique of the Superphosphate Industry," *Superphosphate* 1 (1928): 99. For a discussion of English coprolite use from 1845, see C. C. Hoyer Millar, *Florida, South Carolina, and Canadian Phosphates* (London: Eden Fisher, 1892), 16–17.

4. Speter, "Final Summary of the Research Into the Origin of Superphosphates," *Superphosphate* 8 (1935): 163.

5. *American Agriculturist* 10 (1851): 123–24; *Working Farmer* 3 (1851): 221; 11 (1859): 97.

6. *American Agriculturist* 9 (1850): 220; *Working Farmer* 3 (1851): 91, contains a good description of the processes involved in mining coprolites and manufacturing fertilizers from them.

7. NYSAS, *Transactions* 15 (1851): 326; *Cultivator* n.s., 8 (1851): 59, 249; 14 (1859): 349; *American Agriculturist* 10 (1851): 69, 81–82; *Plough, Loom, and Anvil* 3 (1850–51): 638. For a biographic sketch of Emmons, see W. J. Youmens, *Pioneers of Science in America, Sketches of Their Lives and Scientific Work* (New York: Appleton, 1896).

8. NYSAS, *Transactions* 15 (1851): 326; *Working Farmer* 3 (1851): 75, 108, 221–22, 230, 254; *American Agriculturist* 10 (1851): 123; *Plough, Loom, and Anvil* 3 (1850–51); 638: *Country Gentleman* 14 (1859): 349.

9. *Plough, Loom, and Anvil* 3 (1850–51); 638:

10. *Country Gentleman* 14 (1859): 349; *Working Farmer* 4 (1852): 136. Mapes strongly denied reports that anyone was using New Jersey mineral phosphates to

manufacture superphospates. See also William P. Blake, "Contribution to the Early History of the Industry of Phosphate of Lime in the U.S.," *Transactions of the American Institute of Mining Engineers* 21 (1892): 159.

11. *Working Farmer* 11 (1859): 80. See Blake, "Early History of the Industry," 159, for a discussion of the difficulties encountered in using Hurdtown and Crown Point deposits to produce superphospates. See also USDA, *Report of the Commissioner of Agriculture* (1868): 377–80.

12. Arch Fredric Blakey, *The Florida Phosphate Industry: A History of the Development and Use of a Vital Mineral* (Cambridge: Harvard University Press, 1873), 9. See *Working Farmer* 3 (1851): 230 for a list of forty-six possible sites with mineral phosphates in the United States and Canada.

13. Philip E. Chazal, *The Century in Phosphates and Fertilizers: A Sketch of the South Carolina Phosphate Industry* (Charleston, S.C.: Lucas Richardson, 1904), 32–53. Also see N. A. Pratt, *Ashley River Phosphates* (Philadelphia: N. pub., 1868), 1–5; Francis S. Holmes, *Phosphate Rocks of South Carolina* (Charleston, S.C.: Holmes Book House, 1870), 56–70; USDA, *Report* (1868: 73–77, which mostly follows Pratt; Otto A. Moses, "The Phosphate Deposits of South Carolina," U.S. Department of the Interior, Geological Survey, *Mineral Resources of the United States* (Washington: GPO, 1883), 504–21. For an excellent bibliography on South Carolina phosphates, see William H. Waggaman and Henry W. Eastwood, *Phosphoric Acid, Phosphates, and Phosphatic Fertilizers* (New York: Chemistry Catalog, 1927), 81.

14. Moses, "Phosphate Deposits of South Carolina," 504.

15. Pratt, *Ashley River Phosphates*, 4.

16. Ibid.; Chazal, *Century in Phosphates*, 34–35; Holmes, *Phosphate Rocks of South Carolina*, 56; Charleston, South Carolina, Chamber of Commerce, *Trade and Commerce of the City of Charleston, South Carolina from September 1, 1865 to September 1, 1872* (Charleston Chamber of Commerce, 1873), 46.

17. Pratt, *Ashley River Phosphates*, 4–5; Charles U. Shepard, "The Charleston Phosphates," address delivered before the Medical Association of South Carolina, 1859, 2. Part of Toumey's *Report* is quoted in Chazal, *Century in Phosphates*, 35–36.

18. Holmes, *Phosphate Rocks of South Carolina*, 57.

19. Shepard, "Charleston Phosphates," 2; *American Journal of Science* 22 (1856): 96.

20. Shepard, "Charleston Phosphates," 3.

21. See Chazal, *Century in Phosphates*, 39, 41. Shepard later specifically referred to this "striking resemblance" between some South Carolina phosphatic nodules and the Monk's Island "stone-guano" (which he had named "pyroclasite") in an article in the *American Journal of Science* s. 2, 47 (1869): 339.

22. Hatch letter to *Rural Carolinian* 2 (1870): 357, quoted in Chazal, *Century in Phosphates*, 42.

23. Shepard described this enterprise in a letter to Dr. H. Bayer, 7 November 1868, printed with Shepard, "Charleston Phosphates," 4.

24. George T. Jackson to Charles U. Shepard, Jr., 11 July 1873, quoted in Chazal, *Century in Phosphates*, 41.

25. Chazal, *Century in Phosphates*, 41.

26. Hatch to *Rural Carolinian* 2 (1870): 357, quoted in Chazal, *Century in Phosphates*, 42. See also Shepard to Bayer, 7 November 1868.

27. Shepard, "Charleston Phosphates," 2; Charleston, *Trade and Commerce*, 50.

28. Chazal, *Century in Phosphates*, 43–44.

29. Charleston, *Trade and Commerce* (1873), quoted in Chazal, *Century in Phosphates*, 44.

30. Chazal, *Century in Phosphates*, 44.

31. The author of Pratt's obituary in the *American Fertilizer* 25 (1906): 16, called Pratt the "founder of the modern fertilizer industry," a claim that seems considerably exaggerated.

32. Holmes, *Phosphate Rocks of South Carolina*, 63–64.

33. Pratt, *Ashley River Phosphates*, 12–14; Holmes, *Phosphate Rocks of South Carolina*, 82–83; Charleston, *Trade and Commerce*, 52. See Chazal, *Century in Phosphates*, 44–45, for a slightly different view of this.

34. Pratt, *Ashley River Phosphates*, 17.

35. Ibid.; Holmes, *Phosphate Rocks of South Carolina*, 68, 69, 77. Klett was a member of the Philadelphia fertilizer firm of Potts and Klett. Lewis was a leading Philadelphia chemical entrepreneur (Williams Haynes, *American Chemical Industry*, 1608–1911 [New York: D. Van Nostrand, 1939], 194). The other Philadelphia investors were Jesse E. Smith, Samuel F. Fisher, and Samuel Grant (Charleston, *Trade and Commerce*, 51). All three were merchants (James Gopsill, *Gopsill's Philadelphia Directory for 1867* [Philadelphia: James Gopsill, 1867], s.v.).

36. Figures from Chazal, *Century in Phosphates*, 48.

37. Holmes, *Phosphate Rocks of South Carolina*, 76.

38. Ibid., 26. Cf. Haynes, *American Chemical Industry*, 347, who erroneously reverses the two Charleston firms.

39. Moses, "Phosphate Deposits of South Carolina," 513.

40. U.S. Census Office, Tenth Census (1880), vol. 6, *Report on Cotton Production in the United States* (Washington: GPO, 1884), 468.

41 Ibid., 515; *Wando Mining and Manufacturing Company* (Charleston: N. pub., 1869), 3–4.

42. Chazal, *Century in Phosphates*, 50–51. For a good description of these washers, see Moses, "Phosphate Deposits of South Carolina," 514; and Wando Mining, pamphlet, 4.

43. Pacific Guano Company, *The Pacific Guano Company: Its History; Its Products and Trade; Its Relation to Agriculture* (Cambridge: Riverside Press, 1876), 44. Millar. *Florida, South Carolina, and Canadian Phosphates*, 160.

44. Pratt, *Ashley River Phosphates*, 50–54. According to Holmes, *Phosphate Rocks of South Carolina*, 82, C. C. Coe was the superintendent of the project. Probably he was a relative of the Coe family Bradley had bought out in Boston.

45. Moses, "Phosphate Deposits of South Carolina," 516.

46. Charleston, *Trade and Commerce*, 81–87, and Table B.

47. Chazal, *Century in Phosphates*, 63. Charleston, *Trade and Commerce*, 71, 76–77.

48. Chazal, *Century in Phosphates*, 63–64; Charleston, *Trade and Commerce*, 60. The Pacific Guano Company also operated mines near Jacksonboro for a few years.

49. Charleston, *Trade and Commerce*, 77–79.

50. G. Browne Goode, "A History of the Menhaden," U.S. Commission of Fish and Fisheries, *Report of the Commissioner for 1877* (Washington: GPO, 1879), 488.

51. Charleston Chamber of Commerce, *Trade Review of Charleston* (1873), quoted in Chazal, *Century in Phosphates*, 63, lists six operating plants in 1873: Wando Company, Sulphuric Acid and Superphosphate Company, Pacific Guano Company, Stono Company, Wappo Mills (J. B. Sardy), and Atlantic Company. Of the thirteen companies operating in 1870, listed by Holmes, *Phosphate Rocks of South Carolina*, 78–87, at least seven had northern connections.

52. For John B. Sardy advertisements, see *American Agriculturist* 18 (1859): 378: 19 (1860): 30; 29 (1870): 116.

53. Ranks calculated from U.S. Census Office, *Census of Manufacturers*, 1870–1900.

54. Chazal, *Century in Phosphates*, 70, reproduces a table prepared by Major E. Willis for the Centennial edition of the *News & Courier* showing production and distribution figures for the South Carolina phosphate industry from 1867–1903. See also the table in Moses, "Phosphate Deposits of South Carolina,"518–21. For a chart of production by company, see August Voelcker, *On the Chemical Composition of Phosphatic Minerals Used for Agricultural Purposes* (London: William Clowes & Sons, 1875), 29.

55. For statistics on guano imports, see U.S. Treasury, *Annual Report and Statement on the Commerce and Navigation of the United States*, 1867–1900.

56. USDA, *Report* (1868): 75.

57. Ibid. (1870): 438.

58. Holmes, *Phosphate Rocks of South Carolina*, 45–46; Collings, *Commercial Fertilizers*, 156–58; William H. Waggaman, *Phosphoric Acid, Phosphates, and Phosphatic Fertilizers*, 2nd ed. (New York: Reinholdt, 1952), 40.

59. For example, see Pratt, *Ashley River Phosphates*, 21, 23, 26, who stresses that the deposit was "purely animal in its origin." See also *American Agriculturist* 31 (1872): 20.

60. Charles U. Shepard, *American Journal of Science* s. 2, 47 (1869): 339.

61. According to a table provided by Charles V. Mapes, the New York fertilizer manufacturer, the cost of superphosphate prepared from South Carolina rock phosphates was $25 per ton while the price of superphosphate prepared from bone ranged between $35 and $45 for a product of the same or only slightly higher analysis (*Country Gentleman* 41 [1876]: 644).

62. In the mid-1870s, the Pacific Guano Company still used considerable quantities of Navassa guano, probably because it contained 72 percent phosphate of lime compared to 60 or 70 percent for the South Carolina rock phosphates (Goode, "History of Menhaden," 489).

63. According to George W. Howard, *The Monumental City: Its Past History and Present Resources* (Baltimore: J. D. Ehlers, 1873), 235, Baltimore then used large amounts of bone and slaughterhouse waste brought from the Midwest. Lister Brothers claimed to have been the first to collect buffalo bones in the West. It also imported bones from all parts of the country and from South America (William Ford, *The Industrial Interests of Newark, New Jersey* [New York: Van Arsdale, 1874], 214).

64. *American Fertilizer* 58 (2 June 1923): 44. In 1874, scavengers shipped 5,037 tons of buffalo bones on two Kansas railroads (USDA, *Report* [1875]: 509).

65. John Randolph Bland, *A Review of the Commerce of the City of Baltimore* (Baltimore: The Sun, 1886), 91.

66. See Peter Collier, "Report on Commercial Fertilizers," in Robert H. Thurston, ed., *Reports of the Commissioners of the United States to the International Exhibition Held in Vienna, 1873* (Washington: GPO, 1876), 44; and Bland, *Commerce of Baltimore,* 91.

67. See Connecticut State Board of Agriculture, *Annual Report* (1873): 352–53, and R. A. Clemen, *The American Livestock and Meat Industry* (New York: The Ronald Press, 1923), 367–68.

68. Calculated from information in Bland, *Commerce of Baltimore,* 186–90, nitrogenous wastes cost about $30 per ton, wholesale. Charles V. Mapes lists the price of dried blood or flesh as $70 while unprocessed South Carolina phosphates cost only $10 per ton (*Country Gentleman* 41 [1876]: 644).

69. See Collings, *Commercial Fertilizers,* 4, 27–34; *Cultivator* n.s., 1 (1844): 174; Frank Roy Rutter, *South American Trade of Baltimore* (Baltimore: Johns Hopkins University Press, 1897), 77, 78; MSBA, *Annual Report* 1 (1873): 367–68.

70. *Country Gentleman* 21, (1863): 297.

71. For information on the commercial exploitation of New Jersey greensand marl deposits, see *American Journal of Science* s. 2, 47 (1869): 429; *Squankum Marl as a Fertilizer: Its Uses and Effects* (New York: N. pub., 1868), passim; Pemberton Marl Company, *Marl Manual* (Philadelphia: King & Baird, 1867), passim; *Country Gentleman* 14 (1859): 365; 22 (1863): 9; USDA, *Report* (1868): 381; *American Agriculturist* 18 (1859): 127.

72. Collier, "Commercial Fertilizers," 51.

73. For early articles and advertisements for kainite, see *Country Gentleman* 29 (1867): 395; 38 (1873): 112; Collier, "Commercial Fertilizers," 7; MSBA, *Annual Report* 1 (1873): 369.

74. Goode, "History of Menhaden," 488–89.

75. Joshua Horner began including muriate of potash in his "Maryland Super-phosphate" by 1873 (Howard, *Monumental City,* 273).

76. For a discussion on the German monopoly, see Haynes, *American Chemical Industry,* 350.

77. For two examples of such plants, see U.S. Congress, 53: 2, *Senate Report* 368 (1894) No. 246, 248. Also see U.S. Census Office, *Census of Manufacturers* (1900), 10: 563.

78. J. H. Baker and Brother of New York first advertised this in 1859. See *American Agriculturist* 18 (1859): 127; Samuel W. Johnson, *Essays on Peat, Muck, and Commercial Manures* (Hartford: Brown & Gross, 1859), 169.

79. Bland, *Commerce of Baltimore,* 91.

80. USDA, *Report,* (1873): 273.

81. *Country Gentleman* 35 (1870): 777.

82. *American Farmer* 28 (1869): 326.

83. William Strickland, *Journal of a Tour in the United States of America, 1794–1795,* J. E. Strickland, ed. (New York: New York Historical Society, 1971), 108.

84. *American Farmer* 8 (1826–27): 340.

85. For complaints about various types of frauds, see *Cultivator* n.s., 1 (1844): 246; *Working Farmer* 3 (1851): 75; *American Agriculturist* 16 (1857): 12; 25 (1866): 129; MSBA, *Transactions* 2 (1858): 189; Goodale, *Commercial Manures,* 15; *Country Gentleman* 21 (1863): 297.

86. *Boston Journal of Chemistry,* quoted in Goodale, *Commercial Manures,* 23.

87. MSBA, *Transactions* (1868): 282.

88. Joseph Harris, *Talks on Manures* (New York: Orange Judd, 1883), 325.

89. For advertisements of the "Grafton Mineral Fertilizer," see Davis, Thayer & Company, *The Grafton Mineral Fertilizer and Destroyer of Insects* (Concord, N.H. Republican Press, 1872), passim; and *American Agriculturist* 38 (1872): 119. The editors of the *Agriculturist*, who screened out advertisements that they felt to be "humbugs," allowed the Grafton advertisement even though chemical analysis showed it should be "inert." The editors apparently allowed the advertisement on the strength of testimonials from farmers who had tried the fertilizer. See *American Agriculturist* 31 (1872): 48.

90. Collier, "Commercial Fertilizers," 24–25, 42. See also MSBA, *Transactions* (1870): 197; and *Country Gentleman* 36 (1871): 132, 216, 728.

91. Collier, "Commercial Fertilizers," 42. Even the promoters of the Grafton fertilizer thought the Stevens fertilizer a fraud. See Davis, Thayer & Company, *Grafton Mineral Fertilizer*, 4.

92. See *American Agriculturist* 26 (1867): 48, for comments on problems with "innocent dealers" and "false theories."

93. Fletcher P. Veitch, "Maryland's Early Fertilizer Laws and Her 1st State Agricultural Chemist," *Journal of the Association of Official Agricultural Chemists* 17 (August 1934): 483.

94. Ibid.

95. Johnson, *Essays*, 3. The Connecticut State Agricultural Society fell upon hard times after the publication of Johnson's *Essays* in 1859. Johnson refused to accept reappointment without an increase in salary and the society itself soon folded. Johnson did not resume testing fertilizer until 1869 when he and Wilbur O. Atwater, one of his graduate students at Yale, analyzed sixteen fertilizers for the Connecticut State Board of Agriculture. Atwater and Johnson continued to test fertilizers during the next few years, hoping to use the results to show farmers the need for a state agricultural experiment station, a goal they finally realized in 1876. For an excellent discussion of Johnson's involvement in fertilizer testing and the foundation of the Connecticut experiment station, see Margaret W. Rossiter, *Emergence of Agricultural Science: Justis Liebig and the Americans, 1840–1880* (New Haven: Yale University Press, 1975), 156–70.

96. *American Agriculturist* 21 (1862): 138–39.

97. *Country Gentleman* 38 (1873): 36.

98. George E. White, one of the city's leading fertilizer dealers, complained bitterly about the injustice of the investigation. In a letter to the *Country Gentleman* 38 (1873): 99, he pointed out that since at least three-quarters of his Peruvian guano sales were delivered directly from the Peruvian government warehouse, he could not have adulterated those. Moreover, he would not have had much incentive to adulterate the small quantities sold directly from his store.

99. *Country Gentleman* 38 (1873): 489.

100. Ibid. 39 (1874): 136–37.

101. Veitch, "Maryland's Early Fertilizer Laws," 474, refers to a Massachusetts lime inspection law as early as 1785. This may, however, have been for the inspection of construction lime, not agricultural lime. Maryland also had lime and plaster inspection acts dating from 1833, but these were intended to insure proper weights, not to guarantee analyses.

102. Ibid., 476.

103. Ibid., 477. For an example of Higgins' criticism, see his *Second Report of the State Agricultural Chemist* (Annapolis: Thomas E. Martin, 1852), 67–68; and his *Fourth Report* (1854), 62–64.

104. Veitch, "Maryland's Early Fertilizer Laws," 477.

105. Rutter, *South American Trade*, 41.

106. *American Farmer* 13 (1857–58): 49.

107. Veitch, "Maryland's Early Fertilizer Laws," 480.

108. Rosser H. Taylor, "Fertilizers and Farming in the Southeast, 1840–1950, Part 1, 1840–1900," *North Carolina Historical Review* 30 (1953): 325; and National Fertilizer Association, *The Fertilizer Movement During the Season, 1882–83* ([Baltimore]: The Association, 1883), xv.

109. See USDA, *Report* (1869): 429. Clarence Albert Day, *Farming in Maine, 1860–1940* (Orono, Me.: University Press, 1963), 24–25; Goodale, "Commercial Manures," 18; *American Farmer* 28 (1869): 326.

110. Rossiter, *Emergence of Agricultural Science*, 159.

111. For a good summary of state legislation as it existed in 1883, see National Fertilizer Association, *Fertilizer Movement, 1882–83*, Appendix. For some reason, this compilation omitted the New Hampshire Act of 1872. See also Veitch, "Maryland's Early Fertilizer Laws," 474; Stevenson Whitcomb Fletcher, *Pennsylvania Agriculture and Country Life, 1840–1940* (Harrisburg: Pennsylvania Historical and Museum Commission, 1955), 113–15.

112. *American Farmer* 4 (1875): 443.

113. In the late 1860s and 1870s a number of public and private institutions began publishing extensive fertilizer analyses. Here is a partial list with starting dates: S. W. Johnson, Connecticut State Board of Agriculture, 1868; F. H. Storer, Bussey Institution, Harvard, 1874; William H. Bruckner, *American Manures* (1872), C. A. Groessman, MSBA, 1873; P. Collier, Vermont State Board of Agriculture, 1872; Evan Pugh, Pennsylvania Agricultural College, 1862; New York State Board of Agriculture, 1882; George H. Cook, New Jersey State Board of Agriculture, 1873; Georgia Department of Agriculture, 1874. See Collier, "Commercial Fertilizers," 36–37.

114. Georgia Department of Agriculture, *Analyses and Statistics of Fertilizers Sold in Georgia, During the Season of 1875-76*, Circular No. 26 (Atlanta: Ga. Dept. of Agriculture, 1876), 11.

115. For example, Hobson, Hurtado & Company's 1875 pamphlet, *Rectified Peruvian Guano*, reprinted the entire 1874 Georgia analysis, and Breining and Helfrich of Allentown, Pennsylvania, reprinted the 1870 fertilizer report of the Pennsylvania Agricultural College in its c. 1870 broadside.

116. Rumford Chemical Works, *Eighty Years of Baking Powder History, 1859–1939* (Rumford, R.I.: N. pub., 1939), 8.

117. Rumford Chemical Works, *Wilson's Ammoniated Superphosphate of Lime* (N. pub., [1870]), 16; *The 'Compost-heap' as Prepared with Wilson's Ammoniated Superphosphate of Lime* (Providence: Rumford Press, 1873), 6–7, 51–55.

118. Collier, "Commercial Fertilizers," 13, 26; *American Agriculturist* 34 (1875): 10–11.

119. For example, the *Country Gentleman* 38 (1873): 83, reported that an

eastern Pennsylvania farmers' club had discussed the subject of fertilizer fraud and proposed remedial legislation. See also MSBA, *Transactions* 1 (1872): 177.

120. See for examples, *American Agriculturist* 28 (1869): 326; *Working Farmer* 4 (1852): 77.

121. For instance, in 1877 the National Association of Fertilizer Manufacturers' fertilizer law committee approved a Georgia law, but decided to challenge a recently passed North Carolina law that imposed a $500 tax on each brand of fertilizer sold in the state. The editors of the *American Farmer* agreed with the Manufacturers' Association. See *American Farmer* s. 8, 6 (1877): 147, 219.

122. Henry Bower, *Bower's Complete Manure* (Philadelphia: Chandler, 1867). See also the discussion of fertilizer laws in National Fertilizer Association, *Fertilizer Movement 1882–1883*, 7.

123. See Collings, *Commercial Fertilizers*, 26; Collier, "Report on Commercial Fertilizers," 43; *American Agriculturist* 32 (1873): 50; Goodale, *Commercial Manures*, 18.

124. For examples, see Ford, *Newark*, 215; John Thomas Scharf, *History of Baltimore City and County* (1881; reprint ed., Baltimore: Regional Publishing, 1971), 397, 399.

125. For example, see Henry Bower Chemical Works, *Bower's Complete Manure* (Philadelphia: [Chandler, 1867?]), 11–13.

126. Jesse W. Markham, *The Fertilizer Industry: Study of an Imperfect Market* (Nashville: Vanderbilt University Press, 1958), believes that excessive brand loyalty created some of the irrational demand patterns he observed in the fertilizer industry, but at least initially, this brand loyalty was one of the farmer's few forms of protection.

127. *American Agriculturist* 21 (1862): 139; Goodale, *Commercial Manures*, 17–18.

128. Edwin T. Freedley, *Philadelphia and Its Manufacturers in 1867*, (Philadelphia: Edward Young, 1867), 285.

129. Even the sellers of Peruvian guano were forced to provide guaranteed analysis. See *American Agriculturist* 35 (1876): 335.

130. The *American Agriculturist* 35 (1876): 114, 154, and 171 described the guarantees, printed a copy of the blank agreement, and published a list of consenting companies.

131. The best descriptions of fertilizer plants before 1880 are: The Wando plant in *Wando Mining and Manufacturing*, 6–7; R. W. L. Rasin's Baltimore plant in *Biographical Cyclopaedia of Representative Men of Maryland* (Baltimore: National Biographical Publishing, 1879), 483; Pacific Guano Company in Goode, "History of Menhaden," 489–90. Also see the description of Baugh and Company in chapter 7. For descriptions of the Poole and Hunt mixer, see Campbell Morfit, *A Practical Treatise on Pure Fertilizers* (New York: D. Van Nostrand, 1872), 145; Goode, "History of Menhaden," 490.

132. For a description of superphosphate manufacturing techniques in 1894, see Harvey Wiley, "Mineral Phosphates as Fertilizers," USDA, *Yearbook* (1894): 184–85.

133. Howard, *Monumental City*, 207, 723; Scharf, *History of Baltimore*, 838; *Wood's* [sic] *Baltimore City Directory (1868/69)* (Baltimore: John W. Woods, ca.

1869), 415. Poole and Hunt also produced steam engines and a wide variety of other machinery and cast iron products, including the columns for the Capital dome.

134. Letterhead, 1870, Pacific Guano Company Collection, Baker Library.

135. Gustav Adolph Liebig was born in Austria in 1824, studied pharmacy and chemistry, received his Ph.D. from Carolinum, and came to the United States in 1856 (*Biographical Cyclopaedia of Representative Men of Maryland*, 168).

136. Holmes, *Phosphate Rocks of South Carolina*, advertisements, 3, 4, 7, 13.

137. A. F. Crowell, "Formula Book," 1884–85; Pacific Guano Company Collection, Baker Library.

138. B. G. Firth to A. F. Crowell, 18 April 1878, Pacific Guano Company Collection, Baker Library, requested a new formula to use when his supply of Navassa phosphatic guano ran out.

139. See "Experiment Book" and "Scrapbook to 1883" (which includes analyses made from 1871–1884) in Pacific Guano Company Collection, Baker Library.

140. *American Farmer* s. 8, 4 (1875): 44.

Chapter IX: The Maturation of the Industry

1. This data is conveniently summarized in U.S. Census Office, Twelfth Census [1900], vol. 5, *Agriculture*, pt. 1, (Washington: GPO, 1902), cxi. The reliability of the manufacturing census data is open to question because of unreliable data collection methods. In addition, some fertilizer production was missed since it occurred as by-products of other industries.

2. For 1868 prices, see *American Agriculturist* 27 (1868): 49. Typical prices from 1872 to 1894 are calculated from the average of the three or four leading brands analyzed for the Connecticut State Board of Agriculture. Connecticut State Board of Agriculture, *Annual Report* 1872, 1874, and 1876; and Connecticut Agricultural Experiment Station, *Annual Report* 1879, 1889, and 1894. Census data reveal that prices of the average ton of purchased fertilizer fell from $27 in 1880, to $18 in 1890, and $14 in 1900 (U.S. Census Office, Twelfth Census [1900], vol. 7, *Manufacturers*, pt. 1 [Washington: GPO, n. date], 562). Unfortunately, the census figures may represent the purchase of lower analysis fertilizers rather than less expensive ones.

3. Data from U.S. Census Office, Tenth Census [1880], vol. 3, *Report on the Productions of Agriculture* (Washington: GPO, 1883).

4. Georgia Department of Agriculture, *Analyses and Statistics of Fertilizers Sold in Georgia During the Season of 1875–76* (Atlanta: Ga. Dept. of Agriculture, 1876), 2–7.

5. Based on a sampling of shipping rates in the commercial pages of the *New York Times* in 1874 and 1880.

6. For A. F. Crowell's calculations of the cost of producing "Soluble Pacific Guano" between 1874 and 1885, see his small notebooks and Experiment Book in the Pacific Guano Company Collection in the Baker Library. In an undated letter from J. M. Glidden to Henry A. Barling, both members of the company's board of directors, Glidden congratulated the firm on its profitability, which he indicated was 20 percent, or about $15 per ton (Pacific Guano Company Collection).

7. Production and consumption of fertilizers in North Carolina, South Carolina, Georgia, and Alabama, as a percentage of national totals:

	1870	1880	1890	1900
Production	14	15	28	27
Consumption	—	36	39	32

(Data from U.S. Census Office, *Agriculture and Manufacturers*, 1870–1890.)

8. According to census data, Maryland was the leading producer of fertilizers in 1879, 1889, and 1899. It may also have been in 1869, when census data was noticeably incomplete. Baltimore generally accounted for most of Maryland's production. A comparison of the figures in John Randolph Bland, *A Review of the Commerce of the City of Baltimore* (Baltimore: The Sun, 1886), 91, with the 1880 census data reveals that the eighteen plants in or near Baltimore accounted for three-fourths of the capital and production of the state's forty-eight fertilizer plants. John Thomas Scharf, *History of Baltimore and County* (1881, reprint, Baltimore: Regional Publishing, 1971), 397, cites twenty-seven fertilizer plants in or near Baltimore in 1881, with an 1880 production of 280,000 tons. This tonnage equalled the entire state's production in the 1879 census count (which only reported value of product, not tonnage).

9. George W. Howard, *The Monumental City: Its Past History and Present Resources* (Baltimore: J. D. Ehlers, 1873), 723.

10. The *American Farmer* s. 8, 6 (1877): 219, reported a meeting in Baltimore of the National Association of Fertilizer Manufacturers. The organization's first President was John S. Reese, its second president was Gustavus Ober, and its treasurer and secretary was R. W. L. Rasin, all of Baltimore. Its officer lists included leading manufacturers from Boston, New York, Charleston, and Wilmington, Delaware, but the bulk of the support apparently came from Baltimore and the meetings were held there. Fertilizer manufacturers organized a second association, the National Fertilizer Association, in 1883. Baltimore manufacturers from the Chemical and Fertilizer Exchange of Baltimore apparently constituted the nucleus of this new organization, which lasted until 1887. See National Fertilizer Association, *Fertilizer Movement during the Season, 1882–83* (Baltimore: National Fertilizer Association, 1883), 3.

11. For example, the Pacific Guano Company engaged John S. Reese as its Baltimore agent by 1865. Lister Brothers established a permanent agency in Baltimore in 1872 (*Industries of Maryland: A Descriptive Review of the Manufacturing and Mercantile Industries of the City of Baltimore* [New York: Historical Publishing, 1882], 314.)

12. Scharf, *History of Baltimore*, 391; R. W. L. Rasin, "Growth of the Fertilizer Industry," *American Fertilizer* 1 (1894): 7.

13. Upton, Shaw & Company, *Brighton Bone-Phosphate and Dry Super-Phosphate of Bone Lime* (Boston: Upton, Shaw & Co., 1872), 25.

14. L. B. Darling & Company, *Darling's Animal Fertilizers* (Providence: L. B. Darling and Co., 1883), 6; Baugh & Son's advertisement, *Maryland Directory* (1882): 86; American Agricultural Chemical Company, *Prospectus* (New York: The Company, 1899), 2.

15. Connecticut State Board of Agriculture, *Annual Report* (1869), 72, 74, and

76; and Connecticut Agricultural Experiment Station, *Annual Report* (1879), 89 and 94.

16. See, for example, Georgia Department of Agriculture, *Fertilizers Sold in Georgia, 1875–76*, passim.

17. W. H. Bowker & Company, *Stockbridge Manures* (Boston: The Company, 1877), 8. For a discussion and some criticism of the Stockbridge manures, see W. O. Atwater article in *American Agriculturist* 37 (1878): 291.

18. Bradley Fertilizer Company, *Bradley's Superphosphate of Lime*([Boston]: N. pub., 1881), 47; William Haynes, *Chemical Pioneers*: The Founders of the American Chemical Industry (New York: D. Van Nostrand, 1939), 84; 84; Mapes advertisements, *American Agriculturist* 40 (1881): 119; 45 (1886): cover.

19. The Connecticut Agricultural Experiment Station, *Annual Report* (1879): 57, included analyses of a number of "special fertilizers" by several leading manufacturers. Fertilizers produced by the same company for different crops were remarkably similar while fertilizers produced by different companies for the same crop differed widely. The report concluded that "the manufacturers do not confine themselves very strictly to their formulas, which is a significant indicator of the value they attach to them." Cf. Gilbeart Collings, *Commercial Fertilizers: Their Sources and Use* (Philadelphia: P. Blakiston's Son, 1938), 301, for additional criticism of special fertilizers.

20. Connecticut Agricultural Experiment Station, *Annual Report* (1878): 19; Georgia Department of Agriculture, *Fertilizers Sold in Georgia, 1875–76*, passim; and Virginia State Board of Agriculture, *Report* (1892): 72, 129.

21. For an example of the lists of products many companies offered, See Bradley Fertilizer Company, *Bradley's Superphosphate of Lime* (Boston: The Company, 1881), cover, 42–43, and Lister Brothers, catalogue (N. pub., 1876): cover.

22. U.S. Treasury Department, *Reports on Commerce and Navigation* (1870–1900). Beginning in 1868, the principal importations came from Ganape Island (MSBA, *Annual Report* 1 [1873]: 353). See *American Agriculturist* 34 (1875): 119, 281, 314, for information on changes in consignees and sample advertisements. See *Country Gentleman* 39 (1874): 89, for a discussion of variability of quality after the exhaustion of the Chincha Islands.

23. Rhode Island State Board of Agriculture, *Commercial Fertilizers*, Bulletin No. 1 (July 1897), 5.

24. For an example, see Baugh & Sons, *Baugh & Sons, Manufacturers of Animal Charcoal* (N. pub., [1884]): covers.

25. For instance, the Connecticut State Board of Agriculture, *Annual Report* (1873): 348, reported that the Connecticut Valley Fertilizer Company had distributed in Connecticut 16,000 tons of manure from New York street railway companies and stables.

26. MSBA, *Annual Report* 1 (1873): 63.

27. *American Farmer* s. 8, 4 (1875): 125.

28. For a more detailed description of agricultural practices in the town and its relation to other similar agricultural areas, see Richard A. Wines, "The Nineteenth Century Agricultural Transition in an Eastern Long Island Community," *Agricultural History* 55 (1981): 52, 60–73.

29. It is difficult to assess the accuracy of the data in the manuscript census.

The actual cost of fertilizers purchased appears to be estimates. Moreover, the enumerator missed two farmers whose diaries survive and indicate they purchased fertilizers regularly. The manuscript census also indicated that another farmer purchased no fertilizer, although his diary indicates that he purchased 800 pounds of guano and 1000 pounds of Phelps fertilizers in 1879.

30. The better complete fertilizers sold in New York City for between $40 and $60 per ton ("Market Basket," *American Agriculturist* 36 [1877]: 165).

31. T. H. S. Boyd, *History of Montgomery County, Maryland from Its Earliest Settlement in 1650 to 1879* (1879; reprint ed., Baltimore: Regional Publishing, 1968), 110. U.S. Census Office, 1880, *Agriculture*, 119, however, indicates cost of fertilizer purchased in 1879 as $335,175.

32. *Country Gentleman* 38 (1873): 5.

33. *American Farmer* s. 8, 9 (1879): 89–90.

34. B. W. Arnold, *History of the Tobacco Industry in Virginia from 1860 to 1894* (Baltimore: Johns Hopkins University Press, 1897), 24, 24n.

35. *American Agriculturist* 21 (1862): 139.

36. Ibid. 29 (1870): 405.

37. Ibid. 31 (1872): 19.

38. For example, see ibid. 34 (1875): 479.

39. Ibid., 54.

40. Ibid. 33 (1874): 35. Ibid. 39 (1880): 35–36, stated that home manufacture of fertilizers was only useful in special cases when the farmer was far removed from market towns.

41. Ibid. 33 (1874): 287.

42. Quotes of fertilizer prices began with the March 1876 issue.

43. *American Agriculturist* 33 (1874), was the first volume to use the index heading "fertilizer."

44. Ibid. 36 (1877): 165–66.

45. Ibid. 34 (1875): 140.

46. For example, the *Country Gentleman* 39 (1874), had 12 articles on commercial superphosphates while 38 (1871) had only 3 brief notes on manufactured fertilizers compared with 127 articles on other kinds of manures. The *Country Gentleman* 41 (1876): 644 published a table of fertilizer prices prepared by Charles V. Mapes without any questions although it had been a sharp critic of his father, James J. Mapes.

47. Ibid. 40 (1875): 280.

48. Ibid. 39 (1874): 776.

49. Ibid., 760.

50. Joseph Harris, *Talks on Manures: A Series of Familiar and Practical Talks between the Author and the Deacon, the Doctor, and Other Neighbors, on the Whole Subject of Manures and Fertilizers* (New York: Orange Judd Company, 1883), vii.

51. *American Agriculturist* 33 (1874): 369.

52. Ibid. 34 (1875): 194.

53. E. M. Shelton to J. Harris, 5 May 1876, in Harris, *Talks on Manures*, 350.

54. U.S. Census Bureau, Tenth Census [1880] vol. 3, *Agriculture*. Cf. the county survey of fertilizer use in USDA, *Report* (1876): 122–27. Price differentials calculated from Georgia Department of Agriculture, *Fertilizers Sold in Georgia*,

1875–76, passim, and Connecticut State Board of Agriculture, *Annual Report* (1874):40–49.

55. Rosser H. Taylor, "Fertilizers and Farming in the Southeast, 1840–1950, Part I, 1840–1900," *North Carolina Historical Review* 30 (1953): 311–13. See also Taylor, "The Sale and Application of Commercial Fertilizers in the South Atlantic States to 1900," *Agricultural History* 21 (1947): 48–49; "Commercial Fertilizers in South Carolina," *South Atlantic Quarterly* 21 (1930): 187–89.

56. MSBA, *Annual Report* 1 (1973): 63.

57. Roger L. Ransom and Richard Sutch, *One Kind of Freedom: The Economic Consequences of Emancipation* (Cambridge: Cambridge University Press, 1977), 46–49.

58. Ransom and Sutch, *One Kind of Freedom*, 102–03, 187.

59. Taylor, "Fertilizers in the Southeast," 317.

60. Ibid., 313, 322; *American Farmer* s. 8, 4 (1875): 125; *American Agriculturist* (1875): 15. One of the favorite themes of the American farmer during the first decades after the Civil War was that commercial fertilizers should only be a temporary stopgap until a new system of small farmers could replace the old plantation system. For example, see *American Farmer* s. 8, 4 (1875): 125.

61. See *American Agriculturist* 30 (1871): 20–21 and S. L. Goodale, *Commercial Manures: A Lecture Delivered before the Farmers' Convention Held at Augusta, January, 1869* (N. pub., 1869), 30, for Northern attitudes toward fertilizer use which demonstrate the prevalence of the recycling mentality there. This attitude toward artificial manures had been developing in the North since their introduction twenty years earlier. For instance, the *Working Farmer* 5 (1853): 73, and the *American Agriculturist* 25 (1866): 136, both considered the principal function of artificial manures to be the future production of more and better yard manure to keep the on-the-farm nutrient cycle going.

62. For typical examples of this type of thinking, see Horace Greeley, *What I Know of Farming* (New York: N. pub., 1871), 121; *American Agriculturist* 30 (1871): 20–21; Goodale, *Commercial Manures*, 30.

63. Collings, *Commercial Fertilizers*, 143, 147.

64. U.S. Census Office, Twelfth Census [1900], vol. 10, *Manufacturers*, pt. 4, 565; Collings, *Commercial Fertilizers*, 171.

65. Vincent Sauchelli, *Chemistry and Technology of Fertilizers* (New York: Reinhold Publishing Corporation, 1960), 2; Mehring et al., *Statistics on Fertilizers and Liming Materials in the United States*, USDA, Agricultural Research Service, Statistical Bulletin No. 191 (April 1957), 3.

66. These potassium oxide plants closed after the war when German supplies resumed, but in 1925 potash deposits were discovered in New Mexico (Sauchelli, *Chemistry of Fertilizers*, 368).

67. *New York Times*, 1 January 1888, 8, called plans to form a fertilizer trust unworkable.

68. *National Cyclopaedia of American Biography* (Clifton, N.J.: J. T. White, 1979), 15: 400.

69. Williams Haynes, *American Chemical Industry, 1608–1911* (New York: D. Van Nostrand, 1889) 1: 350.

70. See *New York Times*, 22 April 1906, 9, and 26 May, 1.

71. For attempts to form the U.S. Agricultural Corporation, see *New York Times*, 2 April 1909, 11; 13 April, 10; 26 May, 11; 11 July, pt. 3, 6. See also Haynes, *American Chemical Industry* 1: 351.

Chapter X: A Complex Technological System

1. For an example of Horace Greeley's thinking on this subject, see his *Essays Designed to Elucidate the Science of Political Economy* (1870; reprint ed., New York: Arno Press, 1972), 137–38.

2. Avery O. Craven, *Soil Exhaustion as a Factor in the Agricultural History of Virginia and Maryland, 1606–1860* (Urbana: University of Illinois Press, 1926), 160–61.

Bibliography

Manuscripts

Ideally, this study would be based on extensive research in the records and correspondence of farmers, agricultural experts, and fertilizer merchants and manufacturers. Since little of this type of material has survived, most of the research has necessarily been done in agricultural periodicals and other printed sources.

The only large body of manuscript material relevant to the fertilizer industry is in the Pacific Guano Company Collection at the Baker Library, Harvard University Graduate School of Business Administration. The collection includes fifteen volumes of journals, daybooks, ledgers, accounts, and letterbooks used at the Pacific Guano Company's Woods Hole plant between 1870 and 1889, along with some personal accounts of the Crowell family. Unfortunately, most of the letterbooks and some of the ledgers are nearly illegible, but the daybooks, which cover the period of 1873 to 1883, are usable. In addition to this material, there is a scrapbook with numerous fertilizer analyses made by A. F. Crowell and his recipes for "Soluble Pacific Guano." Another scrapbook includes clippings on the history of the company and its former island possessions up to 1961. In addition, the collection includes small pocket notebooks, legal documents, insurance maps, patents, and even a sample guano bag. The small notebooks kept by A. F. Crowell give the best picture of the day-to-day operations of the Woods Hole plant and the economics of the company's operation. The collection also includes material on fishing operations connected with the company and the Crowell family.

Some manuscript material for three other firms survives. The Eleutherian Mills Historical Library in Greenville, Delaware has four volumes of records for the Baugh & Sons Company and its successor, the Baugh Chemical Company; but these records mainly cover the period from about 1880 to 1930, and contain little relevant to this study. The Maryland Historical Society Library in Baltimore has one box of F. Barreda and Brother Company records (manuscript 2104), but this collection of bills, telegrams, etc., covers only the period from 1854 to 1855 and contains little material on the guano trade, although it does shed some light on the other business interests of the firm. The East Hampton Free Library in East

Hampton, New York, has a diary kept between 1877 and 1878 by William H. Glover, of Southold, Long Island, who divided his time between farming and the operation of a small fish guano plant. This diary contains numerous references to his plant, but does not give a complete picture of its operation.

The *Federal Hill Story* papers (manuscript 71-271) at the Maryland Historical Society contain much useful information about the early days of the Baltimore fertilizer industry. William J. Kelley apparently began a book on the industries, including several fertilizer plants, located in Baltimore's Federal Hill area, but abandoned the project after finishing a rough draft.

Farmers' diaries, journals, and account books have provided useful insights into the way farmers substituted commercial fertilizers for other material they had previously been using. The most useful of these were kept by four farmers in Riverhead, Long Island: the Day Books of Noah Youngs, 1822, 1831–47; the Diaries of Samuel Tuthill, 1863–1920; the Diary of David Halsey Hallock, 1855–90; and the Journal of John T. Downs, 1867–86, all in private Riverhead collections. The Suffolk County Historical Society has the Account Book, 1847–48, of the Sea Serpent Company, a cooperative fishing organization of South Hampton farmers. The Udall family Account Books at the Nassau County Library in East Meadow, New York, contain some records of the family's manure trade with New York City in the 1840s. The East Hampton Free Library has several farm account books, the most useful of which was probably kept by Cornelius Conklin from 1856 to 1859. The Maryland Historical Society also contains a number of farm account books and diaries. The most useful of these are the Bloomsbury Farm Accounts, 1849–62 (manuscript 6); the Farm Journal kept by Harry Dorsey Gough Carroll, 1848–55 (manuscript 219); the John Henry Carroll Account Books, 1814–56 (manuscript 204); the George W. Hyde Farmbook, 1840–88 (manuscript 2429); the Farm Journal, 1852–64, in the William B. Preston Papers (manuscript 978.1); and the Diary of Edward Stabler, 1852 (manuscript 776).

Serials

American Agriculturist. New York, 1842–91.
American Farmer. Baltimore, 1819–80.
American Fertilizer. 1894–96.
American Institute of the City of New York. *Transactions*. 1847–69.
American Journal of Science. 1856–57, 1863.
Baltimore. 1938.
Connecticut Agricultural Experiment Station. *Annual Report*. 1878–94.
Connecticut State Board of Agriculture. *Annual Report*. 1866–76.
Country Gentleman. Albany, 1853–76.
Cultivator. Albany, 1834–60.
De Bow's Review. 1846–60.
Farmer's Magazine. London, 1845–46.
Glen Cove Gazette. Glen Cove, N.Y., 1860–70.
Hunt's Merchants' Magazine. 1839–70.
Massachusetts State Board of Agriculture. *Annual Report*. 1852–77.
[New York] Society for the Promotion of Useful Arts. *Transactions*. 1792–1819.

New York State Agricultural Society. *Transactions.* 1841–71.
New York Times. 1851–1908.
Pennsylvania State Agricultural Society. *Transactions.* 1854.
Plough, Loom, and Anvil. Philadelphia, 1848–56.
Sag Harbor Corrector. Sag Harbor, N.Y., 1822–80.
Scientific American. 1873.
Southern Agriculturist. Charleston, S.C., 1841–46.
Southern Cultivator. Augusta and Athens, Ga., 1843–79.
Southern Planter. Richmond, Va., 1840–75.
Times. London, 1852.
U.S. Congress. *Congressional Globe.* 1852–72.
U.S. Department of Agriculture. *Report of the Commissioner of Agriculture.* 1862–83.
U.S. Department of Agriculture. *Yearbook.* 1894–1902.
U.S. Patent Office. *Annual Report of the Commissioner of Patents, Part 2, Agriculture.* 1837–62.
U.S. Treasury Department. *Annual Report and Statement on the Commerce and Navigation of the United States.* 1851–1902.
Working Farmer. New York, 1849–59.

Other Printed Works

American Agricultural Chemical Company. *Prospectus.* New York: The Company, 1899.
American Fertilizer Company. *Fales' Patent Concentrated Fertilizer.* Boston: The Company, 1869.
American Guano Company. *Prospectus of the American Guano Company.* New York: J. F. Trow, 1855.
——————. *Report of Experiments with American Guano.* New York: W. H. Arthur, 1859.
——————. *Report to the Stockholders of the American Guano Company.* Brooklyn: Jacobs & Brockway, 1857.
Armbruster, Eugene L. *Brooklyn's Eastern District.* Brooklyn: N. pub., 1942.
Association of the Menhaden Oil and Guano Manufacturers of Maine. *The Menhaden Fishery of Maine.* Portland: B. Thurston, 1878.
Atwater, Wilbur O. *On Commercial Fertilizers at Home and Abroad.* Hartford: N. pub., 1874.
——————. "Menhaden and Other Fishes and Their Products as Related to Agriculture." United States Commission of Fish and Fisheries. *Report of the Commissioner for 1877.* Washington: GPO, 1879.
Aulie, Richard P. "The Mineral Theory." *Agricultural History* 48 (1974): 369–82.
Barclay-Allardice, Robert. *Agricultural Tour in the United States and Upper Canada.* Edinburgh: William Blackwood & Sons, 1842.
Barnett, Leroy. "Buffalo Bones in Detroit." *Detroit in Perspective* 2 (1975): 89–96.
Bartlett, Charles L. *Guano: A Treatise on the History, Economy as a Manure, and Modes of Applying Peruvian Guano.* Boston: C. L. Bartlett, 1860.
[Bartlett, Edwin.] *Guano, Its Origin, Properties, and Uses.* New York: Wiley and Putnam, 1845.

Baugh & Sons. *Baugh & Sons: Manufacturers of Animal Charcoal.* N. pub., [1884].
——————. *How to Maintain the Fertility of American Farms and Plantations.*
    Philadelphia: Spangler & Davis, 1866.
Bayles, Richard Mather. *Historical and Descriptive Sketches of Suffolk County, with a
    Historical Outline of Long Island.* 1873. Reprint. Port Washington, N.Y.: Ira J.
    Friedman, 1962.
Bidwell, Percy Wells, and Falconer, John I. *History of Agriculture in the Northern
    United States: 1620–1860.* Washington: N. pub., 1925.
Biographical Cyclopaedia of Representative Men of Maryland and the District of Columbia.
    Baltimore: National Biographical Publishing, 1879.
Blake, J. R. "Guano." *Land We Love* 2 (1867): 261–66.
Blake, William P. "Contribution to the Early History of the Industry of Phosphate
    of Lime in the United States." *Transactions of the American Institute of Mining
    Engineers* 21 (1892): 157–59.
Blakey, Arch Fredric. *The Florida Phosphate Industry: A History of the Development and
    Use of a Vital Mineral.* Cambridge: Harvard University Press, 1973.
Bland, John Randolph. *A Review of the Commerce of the City of Baltimore.* Baltimore:
    The Sun, 1886.
Bommer, George. *New Method Which Teaches How to Make Vegetable Manure.* 2nd
    ed. New York: Redfield & Savage, 1845.
Boston Milling & Manufacturing Company. *Flour of Bone and Other Fertilizers
    Manufactured by the Boston Milling & Manufacturing Co.* Boston: Hollis & Gunn,
    1865.
Bowditch, Henry I. *Public Hygiene in America.* Boston: Little Brown, 1877.
Bower, Henry. *Bower's Complete Manure.* Philadelphia: Chandler, 1867.
Bowker, W. H. & Company. *Stockbridge Manures.* Boston: N. pub., 1877.
Boyd, A. J. "Reminiscence of the Chincha Islands." *Queensland Geographical Jour-
    nal* 8 (1893): 3–12.
Boyd, T. H. S. *History of Montgomery County, Maryland from Its Earliest Settlement in
    1650 to 1879.* 1879. Reprint. Baltimore: Regional Publishing, 1968.
Bradley Fertilizer Company. *Bradley's Super-Phosphate of Lime.* Boston: The Com-
    pany, 1871.
Bradley Fertilizer Company. Standard poster, from 1881 advertising pamphlet.
Brady, Nyle Co. *The Nature and Properties of Soil,* 8th ed. New York: Macmillan,
    1974.
Bromophyte Fertilizer Company of St. Louis, Missouri. St. Louis: The Company, 1871.
Brown Chemical Company. *Powell's Fertilizers.* Baltimore: John D. Lucas, [1883].
Browne, D. J. *The Field Book of Manures; or the American Muck Book.* New York:
    A. O. More, 1853.
Bryan, Edwin H., Jr. *American Polynesia and the Hawaiian Chain.* Honolulu: Tongg
    Publishing, 1942.
Buckman, Harry O., and Brady, Nyle C. *Nature and Properties of Soils,* 7th ed. New
    York: Macmillan, 1960.
Buller, Thomas Wentworth. *Remarks on the Monopoly of Guano.* London: James
    Ridgway, 1852.
Burling, Gilbert. "Long Island Oil Fishery." 1871. Newspaper clipping at East
    Hampton (N.Y.) Free Library.

Burstyn, Harold L. "Chemical Fertilizers and the Industrialization of Agriculture." Paper delivered before the History of Science Society, 26 October 1978. Madison, WI.

Butchers' Slaughtering & Melting Association. *A Description of the Brighton Abattoir and the Animal Fertilizer Made by the Butchers' Slaughtering & Melting Association, Brighton, Massachusetts.* Boston: J. A. Cummings, 1874.

Ceci, Lynn. "Fish Fertilizers: A Native North American Practice." *Science* 188 (1975): 24–30.

Chadwick, Edwin. *Report on the Sanitary Condition of the Laboring Population of Great Britain.* 1842. Reprint. Edinburgh: University Press, 1965.

Chambers, J. D., and Mingay, G.E. *The Agricultural Revolution 1750–1880.* London: B. T. Batsford, 1966.

Charleston, South Carolina, Chamber of Commerce. *Trade and Commerce of the City of Charleston, South Carolina, from September 1, 1865 to September 1, 1872.* Charleston: Chamber of Commerce, 1873.

Chazal, Philip E. *The Century in Phosphates and Fertilizers: A Sketch of the South Carolina Phosphate Industry.* Charleston, S.C.: Lucas-Richardson, 1904.

Chynoweth, James Bennett, and Bruckner, William H. *American Manures; and Farmers' and Planters' Guide.* Philadelphia, 1871.

Clegg, Samuel. *A Practical Treatise on the Manufacture and Distribution of Coal Gas.* London: John Weale, 1841.

Clifford, Harold B. *The Boothbay Region, 1906–1960.* Freeport, Me.: Bond Wheelwright, 1961.

Closson, Mabel H. "Under the Southern Cross." *The Overland Monthly* 21 (1893): 205–16.

Clow, Archibald, and Clow, Nan L. *Chemical Revolution.* London: Batchworth Press, 1952.

Coker, Robert Ervin. "The Fisheries and Guano Industry of Peru." In *Bureau of Fisheries Bulletin,* 333–65. Washington: GPO, 1910.

————. "Peru's Wealth Producing Birds." *National Geographic Magazine* 37 (1920): 537–66.

Collier, Peter. "Report on Commercial Fertilizers." In *Reports of the Commissioners of the United States to the International Exhibition Held at Vienna, 1873.* Washington: GPO, 1876.

Collings, Gilbeart H. *Commercial Fertilizers: Their Sources and Use.* Philadelphia: P. Blakiston's Son, 1938.

Craven, Avery O. *Soil Exhaustion as a Factor in the Agricultural History of Virginia and Maryland, 1606–1860.* 1926. Reprint. Urbana: University of Illinois, 1965.

Cunningham, John T. *Newark.* New Jersey Historical Society, 1966.

Dana, Samuel L. *A Muck Manual for Farmers.* Lowell, Mass.: D. Bixby, 1842.

Danhof, Clarence. *Change in Agriculture: The Northern United States, 1820–1870.* Cambridge: Harvard University Press, 1969.

Darling, L. B. and Company. *Darling's Animal Fertilizers.* Providence: J.A. & R. A. Reid, 1883.

Davis, Thayer & Company. *The Grafton Mineral Fertilizer and Destroyer of Insects.* Concord, N.Y.: The Republican Press Association, 1872.

Davy, Sir Humphry. *Elements of Agricultural Chemistry in a Course of Lectures for the Board of Agriculture.* 1813. New York: Eastburn, Kirk, 1815.

Day, Clarence Albert. *Farming in Maine, 1860–1940.* Orono, Me.: University Press, 1963.

――――――. *A History of Maine Agriculture, 1604–1860.* Orono, Me.: University Press, 1954.

[De Burg, C. B.] C. B. *DeBurg's Superphosphate of Lime.* Boston: N. pub., [1855].

Dictionary of American Biography. New York: C. Scribner's, 1964.

Donahue, Roy L.; Shickluna, John C.; and Robertson, Lynn S. *Soils: An Introduction to Soils and Plant Growth.* 3rd ed. Englewood Cliffs, N.J.: Prentice Hall, 1971.

Duffield, Alexander James. *Peru in the Guano Age.* London: R. Bentley and Son, 1877.

Duffy, John. *A History of Public Health in New York City 1625–1866.* New York: Russell Sage Foundation, 1968.

――――――. *A History of Public Health in New York City 1866–1966.* New York: Russell Sage Foundation, 1974.

Dwight, Timothy. *Travels in New England and New York.* 4 vols. 1821–1822. Reprint. Cambridge: Harvard University Press, 1969.

Earle Phosphate Company. *Earle Phosphate Guano.* Providence: The Company, 1880.

Ellis, Albert F. *Adventures in Coral Seas.* Sidney, Australia: Augus & Robertson, 1936.

Ellis, David M. *Landlords and Farmers in the Hudson-Mohawk Region, 1790–1850.* Ithaca, N.Y.: Cornell University Press, 1946.

Emerson, Gouverneur. *Address Delivered before the Agricultural Society of Kent County, Delaware.* Philadelphia: National Merchant Printer, 1857.

Engelhardt, George W. *Baltimore City, Maryland.* Baltimore: Baltimore Board of Trade, 1895.

Ernle, Rowland E. Prothero. *English Farming Past and Present.* London: Longmans, Green, 1917.

Etiwan Phosphate Company. *Planters' Pocket Companion.* Charleston, S.C.: Lucas & Richardson, [1885].

Falkner, Frederic. *The Farmer's Manual: A Practical Treatise . . . with a Brief Account of All the Most Recent Discoveries in Agricultural Chemistry.* American reprint ed. New York: Appleton, 1843.

――――――. *The Farmer's Treasure: A Practical Treatise on the Nature and Value of Manures.* New York: D. Appleton, 1847.

Fanning, Leonard. "Guano Islands for Sale." *Maryland Historical Magazine* 52 (1957): 346–48.

Farmer's Bone and Fertilizing Company. Philadelphia: N. pub., 1876.

Fisher, Sidney George. *Mount Harmon Diaries of Sidney George Fisher, 1837–1850.* Edited by W. Emerson Wilson. Wilmington: Historical Society of Delaware, 1976.

Flanders, Ralph Betts. *Plantation Slavery in Georgia.* Chapel Hill: University of North Carolina Press, 1933.

Fletcher, Stevenson Whitcomb. *Pennsylvania Agriculture and Country Life, 1640–1840.* Harrisburg: Pennsylvania Historical and Museum Commission, 1950.

――――――. *Pennsylvania Agriculture and Country Life, 1840–1940.* Harrisburg: Pennsylvania Historical and Museum Commission, 1955.

Ford, William. *The Industrial Interests of Newark, New Jersey.* New York: Van Arsdale, 1874.

Freedley, Edwin T. *Philadelphia and Its Manufacturers in 1857.* Philadelphia: Edward Young, 1859.

—————. *Philadelphia and Its Manufacturers in 1867.* Philadelphia: Edward Young, 1867.

Frézier, Amádée Francois. *A Voyage to the South-Sea, and along the Coasts of Chile and Peru, in the Years 1712, 1713, and 1714.* London: Jonah Bowyer, 1717.

Fuess, Claude Moore. *Daniel Webster.* 2 vols. Boston: Little, Brown, 1930.

Fuller, R. B. "The History and Development of the Mining of Phosphate Rock." In *Manual on Phosphates in Agriculture,* edited by Vincent Sauchelli. Baltimore: Davison Chemical Corp., 1951.

Gabriel, Ralph Henry. *The Evolution of Long Island: A Story of Land and Sea.* New Haven: Yale University Press, 1921.

Gates, Paul W. "Agricultural Change in New York State, 1850–1890." *New York History* 50 (1969): 115–41.

—————. *Agriculture and the Civil War.* New York: Alfred A. Knopf, 1965.

—————. *The Farmer's Age: Agriculture 1815–1860.* New York: Holt, Rinehart and Winston, 1960.

Gaussoin, Eugene. *Memoir on the Island of Navassa (West Indies).* Baltimore: N. pub., 1866.

Georgia Department of Agriculture. *Analyses and Statistics of Fertilizers Sold in Georgia, During the Season of 1875–6.* Circular No. 26. Atlanta, 1876.

[Gibbs, Anthony, and Sons.] *Guano: Its Analysis and Effects Illustrated by the Latest Experiments.* London: W. Clowes & Sons, 1843.

Glidden, Donald Emerson. "The Story of Swan Island." Manuscript, Pacific Guano Company Collection, Baker Library, Harvard University Business School, 1960.

Goodale, Stephen L. *Commercial Manures: A Lecture Delivered before the Farmers' Convention Held at Augusta, January, 1869.* N. pub., [1869].

Goode, G. Brown. "A History of the Menhaden." U.S. Commission of Fish and Fisheries. *Report of the Commissioner for 1877.* Washington: GPO, 1879.

Goode, G. Brown, and Clark, A. Howard. "The Menhaden History." In *The Fisheries and Fishery Industries of the United States.* Washington: GPO, 1887.

Graham, Gerald S. "The Gypsum Trade of the Maritime Provinces: Its Relation to American Diplomacy and Agriculture in the Early Nineteenth Century." *Agricultural History* 12 (1938): 209–23.

Gray, A. N. *Phosphates and Superphosphates.* 2nd ed. London: E. T. Heron, 1944.

Gray, Lewis C. *History of Agriculture in the Southern United States to 1860.* Washington: The Carnegie Institution, 1933.

Greeley, Horace. *Essays Designed to Elucidate the Science of Political Economy.* 1870. Reprint. New York: Arno Press, 1972.

—————. *What I Know of Farming.* New York: N. pub., 1871.

Green, Francis Byron. *History of Boothbay, Southport, and Boothbay Harbor, Maine 1623–1905.* Portland: Loring, Short & Harmon, 1906.

Green, Lawrence George. *Panther Head: The Full Story of the Bird Islands Off the Southern Coasts of Africa, the Men of the Islands, and the Birds in Their Millions.* London: S. Paul, 1955.

Greer, Rob Leon. *The Menhaden Industry of the Atlantic Coast.* Washington: GPO, 1915.

Gruss, L. "The 'Mission' to Ecuador of Judah P. Benjamin." *Louisiana Historical Quarterly* 23 (1940): 162–69.

"A Guano Island." *National Magazine* 3 (1853): 553–56.

Hackworth, Green H. *Digest of International Law.* 8 vols. Washington: GPO, 1940–44.

Hague, James D. "On the Phosphatic Guano Islands of the Pacific Ocean." *American Journal of Science and Arts* 34 (1862): 224–43.

Hannam, John. *The Economy of Waste Manures: A Treatise on the Nature and Use of Neglected Fertilizers.* London: Longman, Brown, Green, and Longmans, 1844.

Harriott, John. *Struggles through Life.* 2 vols. London: N. pub., 1808.

Harris, Joseph. *Talks on Manures: A Series of Familiar and Practical Talks between the Author and the Deacon, the Doctor, and Other Neighbors, on the Whole Subject of Manures and Fertilizers.* New York: Orange Judd Company, 1883.

Harrison, Roger W. *The Menhaden Industry.* U.S. Department of Commerce, Bureau of Fisheries. Washington: GPO, 1931.

Harrison Brothers. *Harrison Brothers & Co.* Philadelphia: Harrison Press, 1885.

Harvard College, Bussey Institution. *Bulletin.* N. pub., 1874.

Hathaway's Fertilizers or Superphosphate of Lime. N. pub., [c. 1857].

Hayami, Jujiro, and Vernon W. Ruttan. *Agriculture Development: An International Perspective.* Baltimore: Johns Hopkins University Press, 1971.

Haynes, Williams. *American Chemical Industry, 1608–1911.* New York: D. Van Nostrand, 1954.

————. *Chemical Pioneers: The Founders of the American Chemical Industry.* New York: D. Van Nostrand, 1939.

————. "History and the Technique of the Superphosphate Industry: The Foundation of the Industry." *Superphosphate* 1 (1928): 93–100, 109–14.

Higgins, James. *Second Report of the State Agriculturist Chemist.* Annapolis, Md.: Thomas E. Martin, 1852.

[Hobson, Hurtado & Company?]. *Rectified Peruvian Guano.* [1875].

Hollander, J. H. *The Financial History of Baltimore.* Baltimore: Johns Hopkins University Press, 1889.

Holmes, Francis S. *Phosphate Rocks of South Carolina . . . with a History of Their Discovery and Development.* Charleston, S.C.: Holmes' Book House, 1870.

Hunt, Shane J. *Growth and Guano in Nineteenth Century Peru.* Princeton: Princeton University Press, 1973.

Hutchinson, George Evelyn. "The Biogeochemistry of Vertebrate Excretion." *Bulletin of the American Museum of Natural History* 96 (1950): 1–554.

Industries of Maryland: A Descriptive Review of the Manufacturing and Mercantile Industries of the City of Baltimore. New York: Historical Publishing, 1882.

Jacob, K. D. "Predecessors of Superphosphates" and "History and Status of the Superphosphate Industry." In *Super Phosphate: Its History, Chemistry, and Manufacture,* 8–94. U.S. Department of Agriculture, Agricultural Research Service and Tennessee Valley Authority. Washington: GPO, 1964.

James, Henry Francis. *The Agricultural Industry of Southeastern Pennsylvania.* Philadelphia: N. pub., 1928.

Johnson, Culhbert William. *On Fertilizers.* London: Ridgeway, Piccadilly, 1839.

Johnson, Samuel W. *Essays on Peat, Muck, and Commercial Manures.* Hartford: Brown & Gross, 1859.

Johnston, James F. W. *Lectures on the Applications of Chemistry and Geology to Agriculture.* New York: Wiley & Putnam, 1844.

——————. *Notes on North America: Agricultural, Economical, and Social.* 2 vols. Edinburgh: Blackwood, 1851.

Jordan, Weymouth T. "The Peruvian Guano Gospel in the Old South." *Agricultural History* 24 (1950): 211–21.

Juan, Jorge, and Ulloa, Antonio de. *A Voyage to South America.* 1748. Translated by John Adams, 1806. Abridged edition. New York: Alfred A. Knopf, 1964.

Julien, Alexis A. "On Metabrashite, Zueglite, Ornithite, and Other Minerals of the Key of Sombrero." *American Journal of Science* n.s., 40 (1865): 367–79.

Kalm, Pehr. *Travels into North America.* Translated by J. R. Foster. 3 vols. Warrington and London: N. pub., 1770–71.

Kreps, Theodore J. *The Economics of the Sulfuric Acid Industry.* Stanford: Stanford University Press, 1938.

Lamer, Mirko. *The World Fertilizer Economy.* Stanford: Stanford University Press, 1957.

La Rochefoucauld-Liancourt, Duc de. *Travels through the United States of North America.* 2 vols. London: N. pub., 1799.

Leary, Peter J. *Essex County, N.J., Illustrated.* Newark, N.J.: Hardham, 1897.

Lee, Daniel. "Treatise on the Relation of Peruvian Guano to American Agriculture." In *Peruvian Guano Trade: Statements and Documents in Relation to the Bill Reported by a Select Committee of the House of Representatives, 31st of July, 1854, Imposing a Sliding Scale of Duty on the Importation of Peruvian Guano into the United States,* 22–32. Washington: W. H. Moore, 1854.

L'Hommedieu, Ezra. "Communications Made to the Society, Relative to Manures." NYSPUA, *Transactions* pt. 1, 1 (1792): 63–76.

——————. "Experiments Made by Manuring Land with Seaweed Taken Directly from Creeks and with Shells." NYSPUA, *Transactions* 1, pt. 2 (1794): 99–102.

——————. "Observations on Manures." NYSPUA, *Transactions* 2nd ed., 1 (1801): 231–42.

Liebig, Justis. *Organic Chemistry in Its Applications to Agriculture and Physiology.* 2nd American edition. Cambridge: John Owen, 1841.

Lister Brothers. *Commercial Catalogue.* N. pub., 1876.

Livingston, Chancellor. "Experiments and Observations on Calcarious and Gypsious Earths." NYSPUA, *Transactions* 1, pt. 1 (1792): 25–62.

Lockeretz, William; Shearer, Georgia; and Kohl, Daniel H. "Organic Farming in the Corn Belt." *Science* 211 (1981): 540–47.

Lodi Manufacturing Company. *An Act to Incorporate the Lodi Manufacturing Company: For Purposes of Agriculture, Passed 6th February 1840: With Accompanying Remarks and Documents.* New York: H. Cassidy, 1840.

——————. *New and Improved Poudrette.* New York: N. pub., 1850.

[Long Island Fish Guano & Oil Works.] *Patent Fish Guano.* N. pub., 1858.

Loudon, John Claudius. *An Encyclopaedia of Agriculture.* London: Longman, Hurst, Rees, Orne, Brown, and Green, 1825.

McNall, Neil Adams. *An Agricultural History of the Genessee Valley, 1790–1860.* Philadelphia: University of Pennsylvania Press, 1952.

Mapes, James J. *Mapes' Nitrogenized Super-Phosphate of Lime.* N. pub., [1859]. Hagley Museum and Library, Wilmington, Del.

Markham, Jesse W. *The Fertilizer Industry: Study of an Imperfect Market.* Nashville: Vanderbilt University Press, 1958.

Marti, Donald B. "Agricultural Journalism and the Diffusion of Knowledge: The First Half Century in America." *Agricultural History* 45 (1980): 28–37.

Maryland Directory. Baltimore: J. Frank Lewis, 1882.

Maryland Fertilizing and Manufacturing Company. *Maryland Fertilizer and Manufacturing Company.* Baltimore: J. Murphy, [1881].

Maryland State Agricultural Chemist. *Second Report of James Higgins, M.D., State Agricultural Chemist to the House of Delegates of Maryland.* Annapolis: Thomas E. Martin, 1852.

Mather, Eugene, and Hart, John Fraser. "The Geography of Manure." *Land Economics* 32 (1956): 25–38.

Mehring, A. L.; Adams, J. Richard; and Jacob, K. D. *Statistics on Fertilizers and Liming Materials in the United States.* USDA, Agricultural Research Service, Statistical Bulletin No. 191. N. pub., 1957.

Metropolitan Board of Works. *The Agricultural Value of the Sewage of London.* London: Edward Stanford, 1865.

Millar, C. C. Hoyer. *Florida, South Carolina, and Canadian Phosphates.* London: Eden Fisher, 1892.

Minor, D. K. *Poudrette as a Manure or Fertilizer in Comparison with Other Manures.* N. pub., [1844].

Moore, John Basset. *Digest of International Law.* 8 vols. Washington: GPO, 1906.

Morfit, Campbell. "On Colombian Guano; and Certain Peculiarities in the Chemical Behavior of 'Bone Phosphate of Lime.' " *Journal of the Franklin Institute* 60 (1855): 325–29.

—————. *A Practical Treatise on Pure Fertilizers.* New York: D. Van Nostrand, 1872.

Moses, Otto A. "The Phosphate Deposits of South Carolina." In *Mineral Resources of the United States, 1882.* U.S. Department of Interior, Geological Survey. Washington: GPO, 1883.

Murphy, Robert C. *Bird Islands of Peru.* New York: G. P. Putnam's Sons, 1925.

Murphy, Robert C., and Murphy, Grace E. Barstow. "Peru Profits from Sea Fowl." *National Geographic Magazine* 145 (1959): 395–413.

Muspratt, James Sheridan. *Chemistry, Theoretical, Practical, and Analytical as Applied and Relating to the Arts and Manufacturers.* "Supplementary Matter" by Eben Horsford. 2 vols. London: William MacKenzie, 1860.

National Cyclopaedia of American Biography. Clifton, N.J.: J. T. White, 1979.

National Fertilizer Association [1883–87]. *The Fertilizer Movement during the Season, 1882–83.* Compiled by A. De Ghequier. [Baltimore]: The Association, 1883.

National Fertilizer Association [1925–Present]. *Fertilizer Consumption in the United States.* Compiled by Herbert Willett. Washington: The Association, 1937.

—————. *The Significance of the Word "Guano" in Fertilizer Terminology.* Prepared by Charles J. Brand. Washington: The Association, 1931.

Navassa Phosphate Company, *Navassa Phosphate Company.* Baltimore, 1864.

Nesbit, John Collis. *On Agricultural Chemistry, and the Nature and Properties of Peruvian Guano.* N. pub., 1856.

————. *The History and Properties of the Different Varieties of Natural Guanos.* London: Rogerson & Tuxford, 1859.

————. *On Peruvian Guano: Its History, Composition, and Fertilizing Qualities.* London: Longman, 1852.

Newark Board of Trade. *Newark, New Jersey, Illustrated: A Souvenir of the City and Its Numerous Industries.* Newark, N.J.: Wm. A. Baker, 1893.

New York City, Common Council. *Minutes of the Common Council, 1784–1831.* 19 vols. New York: City of New York, 1917.

New York State, Secretary of State. *Census of the State of New York for 1855.* Albany, N.Y.: C. Van Benthuysen, 1857.

————. *Census of the State of New York for 1865.* Albany, N.Y.: C. Van Benthuysen, 1867.

————. *Census of the State of New York for 1875.* Albany, N.Y.: Weid, Parsons, 1877.

Nichols, Roy F. *Advance Agents of American Destiny.* Philadelphia: University of Pennsylvania Press, 1956.

Norrington, Charles & Company. *Observations on Superphosphate of Lime Manufactured by Charles Norrington & Co.* Plymouth, England, N. pub., [c. 1860].

Odum, Howard T. *Environment, Power, and Society.* New York: Wiley-Interscience, 1971.

Olmstead, Frederic L. *A Journey in the Seaboard Slave States.* New York: Mason Brothers, 1856.

Owens, Hamilton. *Baltimore on the Chesapeake.* Garden City, N.Y.: Doubleday, Doran, 1941.

Pacific Guano Company. *The Pacific Guano Company: Its History; Its Products and Trade: Its Relation to Agriculture.* Cambridge: Riverside Press, 1876.

Parish, Percy, and Oglivie, A. *Calcium Superphosphate and Compound Fertilizers: Their Chemistry and Manufacture.* London: N. pub., 1939.

Parker, Gannett & Osgood. *Coe's No. 1 Improved Superphosphate of Lime.* Boston: Hollis & Gunn, 1861.

Parkinson, Richard. *A Tour of America in 1798, 1799, and 1800.* London: J. Harding, 1805.

Passaic Agricultural Chemical Works. *Lister Brothers' Standard Fertilizers.* N. pub., 1878.

Payne, John H., Jr. "Fertilizer Manufacture." In *Phosphorus and Its Compounds,* edited by John Van Wazer, New York: Interscience Publishers, 1961.

Peck, George Washington. *Melbourne and the Chincha Islands.* New York: C. Scribner, 1854.

Pemberton, Marl Company. *Marl Manual.* Philadelphia: King & Baird, 1867.

Peters, Richard. *Agricultural Inquiries on Plaster of Paris.* Philadelphia: Cist and Marklane, 1797.

————. "On Gypsum." Pennsylvania Society for Promoting Agriculture. *Memoirs* 1 (1808): 156–75.

Philadelphia Guano Company. *Colombian Guano, Brought from the Guano Islands in the Caribbean Sea Belonging to the Republic of Venezuela.* Philadelphia: James H. Bryson, 1856.

Phipson, T. L. "On Sombrerite." *Journal of the Chemical Society* (London) 15 (1862): 277.

Pittman, G. A. *Nauru: The Phosphate Island.* London: Longmans, 1959.

Plan for the Equitable Settlement of the Guano Question. London: N. pub., [1850s].

Pratt, N. A. *Ashley River Phosphates.* Philadelphia: Inquirer Book & Job Printing, 1868.

Ransom, Roger L., and Sutch, Richard. *One Kind of Freedom: The Economic Consequences of Emancipation.* Cambridge: Cambridge University Press, 1977.

Rasin, R. W. L. "Growth of the Fertilizer Industry." *American Fertilizer* 1 (1894): 5–8.

Rees, Abraham. *The Cyclopaedia; Or, University Dictionary of Arts, Sciences, and Literature.* 41 vols. 1st American edition. Philadelphia: Samuel F. Bradford, 1805–1825.

Reese, John S. & Company. *John S. Reese & Co.: General Agents for Soluble Pacific Guano in the Southern States.* Gettysburg, Pa.: N. pub., 1866.

Rhode Island State Board of Agriculture. *Licensed Fertilizers Sampled by the Chemist of the R. I. State Board of Agriculture.* Bulletin No. 2. N. pub., 1897.

Richards, William. *A Practical Treatise on the Manufacture and Distribution of Coal Gas.* London: E & F. N. Spun, 1877.

Robinson, Solon. *Guano: A Treatise of Practical Information for Farmers.* New York: N. pub., 1852.

Rossiter, Margaret W. *Emergence of Agricultural Science: Justus Liebig and the Americans, 1840–1880.* New Haven: Yale University Press, 1975.

——————. "The Organization of Agricultural Improvement in the United States, 1785–1865." In *The Pursuit of Knowledge in the Early American Republic,* edited by Alexandra Oelson and Sanborn C. Brown. Baltimore: Johns Hopkins University Press, 1976.

——————. "The Organization of the Agricultural Sciences." In *The Organization of Knowledge in Modern America, 1860–1920,* edited by Alexandra Oelson and John Voss. Baltimore: Johns Hopkins University Press, 1976.

Rubin, Julius. "The Limits of Agricultural Progress in the Nineteenth-Century South." *Agricultural History* 49 (1975): 362–73.

Ruffin, Edmund. *An Essay on Calcareous Manures.* 1832. Reprint. Cambridge: Harvard University Press, 1961.

Rumford Chemical Works. *The 'Compost-heap': As Prepared with Wilson's Ammoniated Superphosphate of Lime.* Providence, R.I.: Rumford Press, 1873.

——————. *Eighty Years of Baking Powder History, 1859–1939.* Rumford, R.I.: N. pub., 1939.

——————. *Wilson's Ammoniated Superphosphate of Lime.* N. pub., [1870].

Russell, Robert. *North America: Its Agriculture and Climate.* Edinburgh: Adam & Charles Black, 1857.

Russell, Sir E. John. *A History of Agricultural Science in Great Britain, 1620–1954.* London: Allen & Unwin, 1966.

Ruttan, Vernon; Binswanger, Hans P.; and Hayami, Yujiro. *Induced Innovation in Agriculture.* Paper Delivered at Fifth World Congress of the International Economic Association, 29 August 1977, Tokyo.

Rutter, Frank Roy. *South American Trade of Baltimore.* Baltimore: Johns Hopkins University Press, 1897.

Sanitary and Fertilizer Company of the United States. Part 3. Estimated Profits from the Adoption of this System. Philadelphia: Burk & McFetridge, [1880].

Sauchelli, Vincent, ed. *Chemistry and Technology of Fertilizers.* New York: Reinhold Publishing, 1960.

Scharf, John Thomas. *History of Baltimore City and County.* 1881. Reprint. Baltimore: Regional Publishing, 1971.

Schmidt, Hubert G. *Agriculture in New Jersey: A Three-Hundred Year History.* New Brunswick, N.J.: Rutgers University Press, 1973.

Shannon, Fred A. *The Farmer's Last Frontier: Agriculture, 1860–1897.* New York: Farrar & Rinehart, 1945.

Shattuck, Lemuel. *Report of the Sanitary Commission of Massachusetts, 1850.* Reprint. Cambridge: Harvard University Press, 1948.

Shaw, William H. *History of Essex and Hudson Counties, New Jersey.* Philadelphia: Everts & Peck, 1884.

Shepard, Charles U. "The Charleston Phosphates." Address before Medical Association of the State of South Carolina, 1859. In *Ashley River Phosphates* by N. A. Pratt. Philadelphia: Inquirer Book & Job Printing, 1868.

Sheppard, J. H. *A Practical Treatise on the Use of Peruvian and Ichaboe African Guano: Cheapest Manure in the World.* 2nd edition. London: Simpka, Marshal, 1844.

Sheridan, Richard C. "Chemical Fertilizers in Southern Agriculture." *Agricultural History* (1979).

Smiley, Charles Wesley. "The Extent of the Use of Fish Guano as a Fertilizer." In *Report of the Commissioner of Fish and Fisheries for 1881.* Washington: GPO, 1884.

Spafford, H. G. *A Gazetteer of the State of New York.* Albany, N.Y.: H. C. Southwick, 1813.

Speter, Max. "Final Summary of the Research into the Origin of Superphosphate." *Superphosphate* 8 (1935): 141–51, 161–68, 181–90.

Squankum Marl as a Fertilizer: Its Uses and Effects. New York: N. pub., 1868.

Stanton, William. *The Great United States Exploring Expedition of 1838–1842.* Berkeley: University of California Press, 1975.

Stark, Charles R. *Groton, Connecticut: 1705–1905.* Stonington, Conn.: Palmer Press, 1922.

Starr, Merritt. "General Horace Capron, 1804–1885." *Journal of the Illinois State Historical Society* 18 (1925): 250–349.

Stewart, Watt. *Chinese Bondage in Peru: A History of the Chinese Coolie in Peru, 1849–1874.* Durham, N.C.: Duke University Press, 1951.

Stiles, Henry R. *The Civil, Political, Professional, and Ecclesiastical History: And Commercial and Industrial Record of the County of Kings and the City of Brooklyn from 1683 to 1884.* 2 vols. New York: N. pub., 1884.

Strickland, William. *Journal of a Tour in the United States of America, 1794–1795.* New York: New York Historical Society, 1971.

Swem, Earl Gregg. *An Analysis of Ruffin's Farmers' Register with a Bibliography of Edmund Ruffin.* Richmond: Superintendent of Public Printing, 1919.

Swift & White. *Swift & White, Manufacturers of and Dealers in Fertilizers of All Descriptions.* New York: Robert Malcolm, [1876].

Tarr, Joel A. "From City to Farm: Urban Wastes and the Farmers." *Agricultural History* 49 (1975): 598–612.

Taylor, Rosser H. "Commercial Fertilizers in South Carolina." *South Atlantic Quarterly* 21 (1930): 179–89.

————. "Fertilizers and Farming in the Southeast, 1840–1850, Part I, 1840–1900." *North Carolina Historical Review* 30 (1953): 305–28.

————. "The Sale and Application of Commercial Fertilizers in the South Atlantic States to 1900." *Agricultural History* 21 (1947): 46–52.

Taylor, William J. "Investigation on the Rock Guano from the Islands of the Caribbean Sea." *Proceedings of the Academy of Natural Sciences of Philadelphia* (The Academy, March 1857): 91–100.

Teschemacher, James Edward. *Essay on Guano*. Boston: A. D. Phelps, 1845.

Thompson, F. M. L. "The Second Agricultural Revolution, 1815–1880." *Economic History Review* s. 2., 21 (1968): 62–77.

Tisdale, Samuel L., and Nelson, Werner L. *Soil Fertility and Fertilizers*. 2nd edition. New York: Macmillan, 1966.

Tygert, J. E. & Company. *Star Bone Phosphate*. Philadelphia: A. T. Zeising, 1876.

U.S. Census Office, Eighth Census [1860]. *Agriculture of the United States, 1860*. Washington: GPO, 1864.

————. *Manufacturers of the United States in 1860*. Washington: GPO, 1865.

U.S. Census Office, Ninth Census [1870]. *Statistics on Wealth and Industry of the United States*. Washington: GPO, 1872.

U.S. Census Office, Tenth Census [1880]. *Report on the Manufacturers of the United States at the Tenth Census*. Washington: GPO, 1883.

————. Vol. 3. *Report on the Productions of Agriculture as Returned at the Tenth Census* [1880]. Washington: GPO, 1883.

————. Vol. 6. *Report on Cotton Production in the United States*. Washington: GPO, 1884.

U.S. Census Office, Eleventh Census [1890]. Vol. 4. *Manufacturers*. Part 1. Washington: GPO, 1895.

————. Vol. 5. *Agriculture*. Pt. 1. Washington: GPO, 1895.

U.S. Census Office, Twelfth Census [1900]. Vol. 5. *Agriculture*. Pt. 1. Washington: GPO, 1902.

————. Vol. 7. *Manufacturers*. Pts. 1 and 4. Washington: GPO, 1902.

U.S. Congress, 23: 2. *House Executive Document* No. 105. 1835.

U.S. Congress, 31: 1. *Senate Executive Document* No. 38. 1850.

————. *Senate Executive Document* No. 59. 1850.

————. *Senate Executive Dcument* No. 80. 1850.

U.S. Congress, 32: 1. *Senate Executive Document* No. 109. 1852.

U.S. Congress, 33: 1. *House Executive Document* No. 70. 1854.

————. *House Report* No. 347. 1854.

————. *Senate Miscellaneous Document* No. 18. 1854.

U.S. Congress, 33: 2. *Senate Executive Document* No. 31. 1855.

U.S. Congress, 34: 1. *Senate Miscellaneous Document* No. 60. 1856.

U.S. Congress, 34: 3. *Senate Executive Document* No. 25. 1857.

U.S. Congress, 34: 3. *Senate Report* No. 397. 1856.

U.S. Congress, 35: 1. *Senate Executive Document* No. 69. 1858.

————. *Senate Miscellaneous Document* No. 267. 1858.

————. *Senate Report* No. 307. 1858.

U.S. Congress, 35: 2. *Senate Executive Document* No. 25. 1859

—————. *Senate Report* No. 379. 1859.

U.S. Congress, 36: 1. *Senate Executive Document* No. 37. 1860.

—————. *Senate Report* No. 280. 1860.

U.S. Congress, 36: 2. *Senate Executive Document* No. 10. 1861.

U.S. Congress, 40: 2. *Senate Executive Document* No. 38. 1868.

U.S. Congress, 40: 3. *House Miscellaneous Document* No. 10. 1868.

U.S. Congress, 48: 1. *House Report* No. 252. 1881.

U.S. Congress, 50: 2. *House Report* No. 3878. 1889.

U.S. Congress, 51: 2. *House Report* No. 4040. 1891.

U.S. Department of Agriculture. *Atlas of American Agriculture*. Washington: GPO, 1935.

U.S. Department of Interior, Geological Survey. *Mineral Resources of the United States*, 1882. Washington: GPO, 1883.

United States Guano Company. *Report to the Stockholders*. New York: The Company, 1859.

Upton, Shaw & Company. *Brighton Bone-Phosphate and Dry Super-Phosphate of Bone Lime*. 1872.

Ure, Andrew. *A Dictionary of Arts, Manufactures & Mines*. New York: D. Appleton, 1847.

—————. *A Dictionary of Chemistry*. Philadelphia: Desilver, 1831.

—————. *Ure's Dictionary of Arts, Manufactures, and Mines*. Edited by Robert Hunt. London: Longman, Green, Longman & Roberts, 1861.

Van Deusen, Glyndon G. *William Henry Seward*. New York: Oxford University Press, 1967.

Veitch, P. Fletcher. "Maryland's Early Fertilizer Laws and Her 1st State Agricultural Chemist." *Journal of the Association of Official Agricultural Chemists* 17 (1934): 474–83.

Voelcker, Augustus. *On the Chemical Composition of Phosphatic Minerals Used for Agricultural Purposes*. London: William Clowes & Sons, 1875.

—————. *On Phosphatic Guanos*. London: W. Clowes & Sons, 1876.

Waggaman, William H. *Phosphoric Acid, Phosphates, and Phosphatic Fertilizers*. 2nd edition. New York: Reinholdt, 1952.

Waggaman, William H., and Eastwood, Henry W. *Phosphoric Acid, Phosphates, and Phosphatic Fertilizers*. New York: Chemical Catalogue, 1927.

Wando Mining and Manufacturing Company. Charleston: N. pub., [c. 1869].

Waring, George E., Jr. *Draining for Profit; and Draining for Health*. New York: Orange Judd, 1867.

—————. *The Elements of Agriculture: A Book for Young Farmers*. Montpelier, Vt.: Samuel M. Walton, 1855.

Webb, William H. *Guano from Baker's and Jarvis Islands in the Pacific Ocean; Imported by William H. Webb*. New York: Slote & James, [1862].

Weber, Gustavus A. *The Bureau of Chemistry and Soils: Its History, Activities, and Organization*. Baltimore: Johns Hopkins University Press, 1928.

Wedbee, T. Courtenay J. *The Port of Baltimore in the Making 1828 to 1878*. Baltimore: F. Bowie Smith and Son, 1953.

Werlich, David P. *Peru: A Short History*. Carbondale, Ill.: Southern Illinois University Press, 1978.

Wheeler, Homer J. *Manure and Fertilizer*. New York: Macmillan, 1914.

Wiley, Harvey W. "Mineral Phosphates as Fertilizers." United States Department of Agriculture, *Yearbook* (N. pub., 1894): 177–92.

Williams, Clark & Company, *Williams, Clark & Co., Manufacturers of High Grade Bone Fertilizers.* New York: William J. Schaufels, 1886.

Wines, Richard A. "The Nineteenth Century Agricultural Transition in an Eastern Long Island Community." *Agricultural History* 55 (1981): 50–63.

Wiser, Vivian. "Improving Maryland's Agriculture, 1840–1860." *Maryland Historical Magazine* 62 (1962): 105–32.

Wood & Grant. *Phosphatic Guano, from Sombrero Island, West Indies; The Richest Deposit of Phosphate of Lime Known to the World, Containing above 80 Per Cent of Bone Phosphate of Lime.* New York: N. pub., [1857].

Woodward, Carl R. *The Development of Agriculture in New Jersey: 1640–1880.* New Brunswick, N.J.: Rutgers University Press, 1927.

Wright, Gavin. *The Political Economy of the Cotton South: Households, Markets, and Wealth in the Nineteenth Century.* New York: W. W. Norton, 1978.

Wyatt, Francis. *The Phosphates of America: Where and How They Occur; How They Are Mined; and What They Cost.* New York: Scientific Publishing, 1891.

Youmans, W. J. *Pioneers of Science in America, Sketches of Their Lives and Scientific Work.* New York: Appleton, 1896.

Index